I0055822

An Excursion in Diagrammatic Algebra

Turning a sphere from red to blue

KE Series on Knots and Everything — Vol. 48

An Excursion in Diagrammatic Algebra

Turning a sphere from red to blue

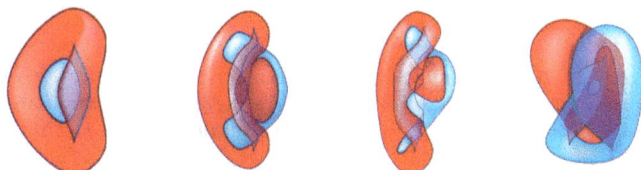

J Scott Carter
University of South Alabama, USA

World Scientific

NEW JERSEY · LONDON · SINGAPORE · BEIJING · SHANGHAI · HONG KONG · TAIPEI · CHENNAI

Published by

World Scientific Publishing Co. Pte. Ltd.

5 Toh Tuck Link, Singapore 596224

USA office: 27 Warren Street, Suite 401-402, Hackensack, NJ 07601

UK office: 57 Shelton Street, Covent Garden, London WC2H 9HE

Library of Congress Cataloging-in-Publication Data
Carter, J. Scott.
    An excursion in diagrammatic algebra : turning a sphere from red to blue / by J. Scott Carter.
    p. cm. -- (Series on knots and everything ; v. 48)
    Includes bibliographical references and index.
    ISBN-13: 978-981-4374-49-1 (hardcover : alk. paper)
    ISBN-10: 981-4374-49-0 (hardcover : alk. paper)
    1. Low-dimensional topology. I. Title.
    QA612.14.C37 2012
    514'.2--dc23

                                          2011034153

British Library Cataloguing-in-Publication Data
A catalogue record for this book is available from the British Library.

Copyright © 2012 by World Scientific Publishing Co. Pte. Ltd.

All rights reserved. This book, or parts thereof, may not be reproduced in any form or by any means, electronic or mechanical, including photocopying, recording or any information storage and retrieval system now known or to be invented, without written permission from the Publisher.

For photocopying of material in this volume, please pay a copying fee through the Copyright Clearance Center, Inc., 222 Rosewood Drive, Danvers, MA 01923, USA. In this case permission to photocopy is not required from the publisher.

Printed in Singapore by Mainland Press Pte Ltd.

Dedicated to

Tom Banchoff, George Francis, and Tony Robbin:
the masters from whom I learned this art

Preface

A 2-dimensional sphere can be turned inside out via a process that allows the sphere to pass through itself, but that keeps tangencies intact. Here I turn the sphere from red to blue, or more precisely from magenta to cyan, but I could reverse the colors and turned it from blue to red, or from orange to green. The color choices are immaterial; this book describes the process.

The result described in the first sentence is over 50 years old. It gives rise to some of the most complicated yet beautiful examples in modern topology. Even the statement seems peculiar: I will have to describe planes of tangency; I will have to describe how tangent planes can become singular; and I will have to explain the precise nature of the sets on which the sphere intersects. Please be patient. There are several key ideas to be developed: surface, sphere, tangent planes, singularities of maps, cusps, folds, and intersection points are among the main ones. Fortunately, all of these ideas can be seen within the world of experience.

Thus, this book is my attempt to explain this example to a lay-public. The example is chosen because of its intrinsic beauty. My version of the sphere eversion builds on the work of others — principally upon the sphere eversion of Froisart and Morin. As I develop the narrative below, I will try to motivate the process and explain the key steps in this construction. Within this preface, I will give an overview of the process.

As I write this, I imagine you to be a college student beginning a pursuit of mathematics or science. You might also be a practicing scientist, computer scientist, or engineer who always appreciated mathematics but did not study it beyond the level of calculus. I imagine you to be curious and to be invigorated by mathematical and scientific imagery. Your interest in mathematics might be akin to my own interest in biology or chemistry. You have an appreciation for the aesthetic of the subject but you have not acquired the technical expertise to be a practitioner.

The measure of the success of this book is how it affects you. Will you learn something in the process of reading it? Will you learn something in the process of looking at the pictures? Will you examine the pictures in details, and determine how they are related? If you have never learned calculus, will this example help you understand the key concept of critical behavior? If you have learned calculus, will you say, "Oh, yeah, I remember that idea?" Will you dust off the calculus book, and look for deeper meaning? Even if the answer to all of these questions is, "No," the book is a success if you come to a greater appreciation of mathematics. Such is my hope.

As a student, I read the works of Gamow, Martin Gardner, and popular writings of Einstein. As an adult, I often read the popular writings of current physicists. As a professional mathematician, my desire is to create similar works for the lay-public that explains the portions of mathematics that I understand well.

―――――――

My immediate tasks are to explain: (a) how the eversion of this book differs from the elegant eversions that already exist; (b) why you should be interested in fully understanding this eversion; (c) the fundamental idea behind this eversion and how I recognized within the process that I made progress.

In regards to the first item, the elegance of the Thurston-Thurston eversion "outside-in" has never been in doubt. It contains the belt trick as an organizing principle. Within the text below, I will give a synopsis of that idea. The Froisart-Morin eversions, the mini-max eversions, and the tobacco-pouch eversions are all beautiful and concise. Their concision is due in no small part to the fact that they are described by a red-blue symmetry. The eversion of this book is noticeably asymmetric. The problem in the symmetric pictures is that at the central moment, the figure is often singular.

Let me give you a metaphor. Perform an experiment with two different objects: a coin and a round potato chip (crisp to the British reader). The chip that I imagine is sold in a can. Hold the coin in your hand with its edge vertical, so that you see the head of the coin. Now slowly turn the coin to tails. In an intermediate stage the coin faced you edge-on. At that stage, the circular boundary of the coin appeared to you as a line segment. That stage was singular: if we conceive of a circle being turned over, then at the singular stage we do not see a circle but a line segment. The situation with the coin simulates a process: clockwise oriented circle, clockwise oriented ellipse, line segment, counter-clockwise oriented ellipse, counter-clockwise oriented circle.

The singular event of the line segment can be decomposed into a less singular situation as follows.

The potato chip does not naturally lie in the plane. It is a saddle shape. Hold the chip and turn it over slowly. I expect that you will see that its boundary circle passes to a figure 8 double loop with one eye of the 8 smaller than the other, the larger loop shrinks until the loops are even tempered, and finally the larger loop becomes the smaller to eventually shrink to nothing. Imagine the boundary of the chip as a 1-dimensional loop. So if the boundary were initially oriented in a clockwise direction, it would become counter-clockwise. There are indeed two singular events in this process: when the figure 8 appears and subsequently disappears, but these events are more general than the circle vanishing into a segment and reappearing over-turned.

The classical eversions are singular in the sense that the turning of the coin is singular. The turning of the chip is less singular. There is more detail that can be gleaned from the shadow of that event than the shadow of the overturned coin. The eversion that I present here is virtually as non-singular as it can be made. Furthermore, at each stage of the eversion I have cut the sphere into slices; that is, I have described each stage as a movie. And successive stages are related via moves-to-movies. Therefore a diligent analysis of the diagrams decomposes the steps into understandable small pieces. This is the main virtue of my approach: *each step can be understood to be built from atomic pieces.*

Consider an algebraic computation. We all have some personal level of algebraic (in)tolerance. One person may look at an equation, and process the solution through a series of mental steps. From the point of view of one who is less algebraically skilled, the first person's computation seems miraculous. In fact, the experienced algebraist may be able to process a sequence of successive steps quickly, whereas the novice needs to see each step written down in order to process it.

In the face of an algebraic computation, I am often the second person. I usually need to see each line of algebraic manipulation worked out explicitly. My algebraic colleagues do not often show me patience. On the other hand, I have trained my own mind to be able to do geometric computations quickly. And those same algebraic colleagues who did not show me patience can be exasperating when they do not see the geometric steps that connect one picture to another. The master of one art often is the novice in another.

The illustrations of the sphere eversion that are presented here are the step-by-step geometric computations that are skipped over or are conglomerated in a computer animation. Clearly, if the computer can perform the computation, as in the case of outside-in or minimax, then the idea must be simple even if the implementation (which requires the computer) is complicated. The result seems miraculous, and we are often overwhelmed by the force of the computer calculation. But the eversion of this book did not use the computer in an essential way. That is not to say that the idea is particularly complicated — quite the opposite. It is designed to illustrate the steps that are often hidden within the classical eversions. No detail is obscured; the difference between successive events is always explicit.

There is a deeper meaning within the calculations. The geometric calculations that are being performed are a geometric encoding of some algebraic calculations. The algebraic notation for those calculations is cumbersome and replete with complication. When translated into diagrams, the algebraic relations appear to be kinematic. Because the diagrams appear to dance on the page when an algebraic identity is translated into geometry, the subject has been dubbed "higher dimensional algebra." The diagrams of the eversion are relationships in that algebraic system, and successive spheres that are illustrated differ by identities among these relationships.

Higher dimensional algebra is an active branch of mathematical research that hopes to relate both to the current state of affairs in physical theories and to help solidify and unify many branches of mathematics. In a nutshell, the underlying principle of higher dimensional algebra is to stop declaring two different things to be equal when what you mean is that there is a natural equivalence between them. Equality is a strict rule: it means that one thing *is* another — a peculiar declaration indeed. But two things are different, so we should compare them, study their differences, and study further the relationships that set them apart or that make them appear to be the same.

A sphere that is red on the outside and blue on the inside is different than one of the opposite colors. An eversion studies how they are related. We can even ask if the eversion here is in substantially different than the other eversions. To answer that question, one needs to have measurements — quantifiable aspects — that can distinguish them. In higher dimensional algebra, we compare the differences between the aspects that measure differences. In Chapter 10, I will discuss aspects of higher dimensional algebra through a physical metaphor. That metaphor is not terribly far from various mathematical models of particle interactions. In fact it is a "baby-model" of particle physics — one that can be used to provide a new language in which the problem of physics (a unification of gravity and quantum mechanics) might be addressed. The subject is way too young to solve this issue, but throughout history, mathematical abstraction has lent itself to addressing problems for which it was not originally intended. And once the abstract had been established, differing areas of mathematics could be used to explain the new abstractions.

There is a fundamental idea that I used in completing this eversion. The double points, triple points, the fold lines, and the quadruple point all interact. At each stage, their interrelations are tracked. For example, the quadruple point separates the red and blue side of the eversion. On the red side, I worked to create a quadruple point. I followed some animations of the Froisart-Morin eversion, but my method never completely coincides with the Froisart-Morin. Near the quadruple point, there must be four triple points, and they must form the vertices of a tetrahedron whose edges are double points; the double point arcs must be the edges of some triangles. The surface is moved into that configuration. Following the quadruple point, it is very easy to cancel one pair of triple points. The remaining pair need to be repositioned. And that repositioning is caused by affecting the fold set, and essentially twisting one end of the immersion a full 360 degrees. On the blue side you will see the top half slowly un-twisting until the triple points can be joined by a triple of arcs of double points.

One can think of an eversion as a type of "quebra-cabaça:" that is a puzzle made of flexible and non-flexible pieces in which some piece is to be apparently unlinked from the main frame. The trick of the toy is to know the correct sequence of hand motions to link or unlink the puzzle. It is usually clear when the puzzle is taken apart and when it is put together. In the eversion, that clarity occurs precisely at the quadruple point. Before the quadruple point, the sphere is red. After it is blue. But just as the puzzle may be solved in a single step, un-solving it can be remarkably tricky. In the eversion of this book, not enough twisting occurred on the red side, so extra twisting has to occur on the blue side.

Nevertheless, progress is measured when the relationships among the fold lines and the double points are configured correctly. In my own solution of the puzzle, I looked alternatively at the chart, the movie, and the decker-set. I looked for moves that could be made, and that put the double points into a more desirable configuration. It is difficult to explain the idea further without delving into the eversion. We will do that shortly.

First acknowledgements are due, and second I will leisurely describe surface, fold, intersections, and relations among interactions.

This book was funded by the National Science Foundation as part research grants DMS-0301095 and DMS-0603926. The opinions expressed here are those of the author and do not reflect the opinions of the National Science Foundation, The University of South Alabama, or anyone else of whom I can think. American taxpayers bought my university this computer and the software that I used to draw the pictures. You have paid for me to discuss this work at conferences. I spent much of my professional time and home life over the past two or three years in the production of this book. Thank you.

Many people deserve more thanks than I can adequately give them. Nonetheless let me try. On a professional level, Masahico Saito, Joachim Rieger, and I developed the mathematical theory of movie moves that are used here. Masahico deserves extra-credit for his patience. Student Sarah Gelginser Brewer suffered through a disorganized stack of rough sketches that outlined the eversion here. My colleagues at the University of South Alabama witnessed the eversion illustrations in their nascent form. I am often insufferably proud of my achievements. They were tolerant. Dr. Cynthia Schneider carefully proof-read a preliminary draft and urged some language changes; many of these were implemented.

My family often looked at the back of my head as I typed or drew the pictures in the corner of the living room. I don't think I neglected them too much over this period, but only their future therapists will be able to confirm that. So thank you to my wife, Huong, and three sons, Albert, Alexander, and Sean. They too witnessed the over-abundance of pride that I am prone to display.

I think that I did most of my daily chores while writing this, but as a writer who sees the steps needed to complete a work, I caution future writers. To complete a book takes long term dedication. One has to reserve blocks of time each and every day. Some other aspects of your life may have to give way. I hope the reader and the causal viewer will see the effort that I brought to the book. Still it is better than digging ditches.

My hope is that the energy and effort that I have given to developing this book will be repaid by the readers' enthusiasms for the subject. Let us begin.

Contents

Chapter 1

A Sphere

A sphere is a 2-dimensional surface. Neglecting subterranean caves and high-rise apartment buildings, you or I could locate the other on this earth by giving the other a pair of numbers: longitude and latitude. I am writing this from $(30°41'10''N, 88°10'59''W)$; thus you can discern that I am writing from Mobile, Alabama a small southern city located close to the Gulf of Mexico.

The *unit sphere* is a subset of a 3-dimensional coordinate space that consists of the solutions, (x, y, z), to the equation

$$x^2 + y^2 + z^2 = 1.$$

It is perfectly round, encompasses a center at the point $(0, 0, 0)$ (known as the *origin* in space), and has a radius of 1. Each of its points is exactly 1 unit away from the origin. It is smooth, polished, perfect, and perfectly thin, more thin than the bubble of soap film that was blown a few moments ago and now is about to dry and vanish.

We adore spheres. The gaseous planets, Jupiter, Neptune, and Saturn (stripped of its rings) are gigantic gods once devoured by Chronos like so many Skittles eaten as you pass the candy counter. The third aisle in the super-market houses a basket of colorful marbled play toys — one dollar each, two dollars for the large ones — screams of joy upon receipt, screams of sorrow upon denial. How fragrant and delicious are those oranges, packed with sticky juices that dribble from your chin. The tennis ball, the soccer ball, the basketball, each gives freedom. "Go outside and play!" Hit the sphere with a stick. Bounce it on the ground. Listen to its impact as it rebounds from the wooden floor in a ball/bell timbre reverberating from the rafters. Spheres numbered 1 through 15 arranged triangularly on green felt table scatter and rebound in perfectly inelastic collisions when smacked by the slightly larger white cue.

But we misapprehend them too. For by *sphere*, I mean that infinitesimally thin layer that encompasses the material within. Neither sun nor
moon is as round, smooth, or homogenous as the unit sphere defined by
the quadratic equation $x^2 + y^2 + z^2 = 1$. The sun, neither solid nor liquid,
erupts in a violent cataclysm. Flames extend into space, magnetic winds
blow and break the signal of urgent phone calls. The man in the moon
manifests the intricate topography of a dry, gray, inhospitable world. A
sphere separates space into two pieces: that which is inside the sphere and
that which is outside. Within the space of 3-dimensional coordinates, the
inside and outside are different. The former is bounded and the latter is
endless.

Each point, (A, B, C), of the unit sphere engenders a plane, $Ax + By + Cz = 1$, that touches the sphere at exactly that point: *a tangent plane*. A
radial ray emanating from the center of the sphere through the point pierces
the sphere and the tangent plane at that point. The ray lies perpendicular
to the tangent plane as a spindle with a cash register receipt affixed. The
tangent plane approximates the sphere in a neighborhood of the tangency
to an extraordinary degree of accuracy. Your play toy lies sullen and still,
tangent to a wooden floor which, in its turn, approximates the surface of
the earth. The collection of tangent planes manifests the 2-dimensional
nature of the sphere. We attempt to flatten our own world by paving
it, or pouring concrete foundations for our houses. There is a stability in
Euclidean geometry: straight lines, flat planes, right angles, and triangular
support. These flat things approximate our world as the tangent plane
approximates the sphere. When we look closely at the spheres of our lives,
we ignore their topography and idealize them in neighborhoods of points
by their tangent planes.

———————

Let us sit across the table eye-to-eye and hold an orange between us.
We both see an orange disk. The light reflects and suggests a spherical
shape, but on a cloudy day, we may not see that reflection. The orange
may appear to each of us as a drably orange disk like the sun partially
obscured by clouds. No matter how you rotate the orange, you and I still
see a disk. Most of what I see is hidden to you, most of what you see is
hidden to me. But there is a feature common to both of our viewpoints:
that circle that forms the profile of the orange.

It is the same circle that delineates the sun in the sky or the full moon on a cloudless night. These orbs, the orange between us, and the toy on the floor all have a circle that encompasses their profile. The profile-defining circle of the orange is the set upon which the tangent planes lie perpendicular to both of our fields of view. I do not see the planes tangent to your side of the orange and you do not see mine. The object between us obscures them. Yet your tangent planes are projected to the same points of my retina as mine are, and vice versa. Except at the delineating profile, two points are projected to one, and the tangency approximates our views.

Chapter 2

Surfaces, Folds, and Cusps

When you or I look at *any* surface, we see it in profile. The *profile* delineates the surface from its surrounding. The profile is the line on which the tangents are turned away from the eye. Take this page, and turn the book so the page appears as a line. You see no print, you read no words, you see no page. The page is a plane of tangency to its own surface. The edge of the page is a rectangle, the set which delineates the page. I look down upon my shadow. The shadow is a projection of my body on the floor. It is defined by its profile.

I look down at my fingers as I pause from typing. The upper profile of my index finger obscures the lower profile of my middle finger, and the middle finger does the same to the ring finger which completely obscures my baby finger. My hand is poised and ready to continue typing. There is a cusp at the junction of the index finger, and I can imagine the line at the knuckle curving below and extending to the lower profile of the next finger.

Venus's statue stands poised on a mantle to my right. I stand to look down upon it from the front. The two folds that outline each breast form a pair of arcs. Each arc appears to end in a pair of cusps, but the folds do not truly end; they continue below the breast and beyond my point of view. Those invisible folds attach her breasts to her body. Most of the folds of our own bodies remain unseen: the fold behind your knee, the crease of your elbow, or the side of your nose as your face is turned. The visual appeal of a human form is accentuated by its cusps and folds. The artist uses these cusps and fold to sketch a form.

The lines that form the shape of the body are the lines upon which the planes tangent to the surface vanish when we look. Mathematical forms are idealized from the lines. The artist, aware of these lines, does not draw where one imagines them to go. The lines below the cusps hide.

————————

When a surface is projected in a usual way to a plane, most points of the surface have neighborhoods which are simulated by their tangent planes. Some points form *folds*: the points at which the tangent plane projects singularly to the field of vision. The folds form a 1-dimensional set that consists of arcs that end in cusps or folds that form closed curves. Each *cusp* is a point at which the tangent plane degenerates to a point upon projection, but the cusp itself has folds emanating from it: one unobstructed from the field of vision, and the other hidden by the remaining surface. The surface, thus projected, is not folded. That Venus is certainly not flat, nor is my dress shirt destroyed when it is ironed; the crease in the sleeve is an artifact of projection of the field of view.

Chapter 3

The Inside and Outside

But we are talking about spheres, and one family of spheres is the subject of this book. So let the discussion return to the unit sphere in space. This point set $\{(x, y, z) : x^2 + y^2 + z^2 = 1\}$ separates space into two pieces, an inside and an outside. And the surface of the sphere is also two sided. (Not every surface that we encounter in this book is two-sided, but that story comes later.) Let us paint the outside of the sphere red and the inside blue. No wait, paint has thickness and our unit sphere is no longer an image of perfection with a pair of thick latex coats upon it. Since the sphere is mathematical, we can pigment the sphere on its two sides: outside red, inside blue, and consider those pigments to be as idealized as the sphere itself.

This book has one purpose: to turn the sphere inside-out in such a way that during the process, every point has a neighborhood which is approximated by a plane tangent at that point and that tangent plane can be thought of as a small flat surface in space. The tangent plane never degenerates to a line or to a point. I will say that under these circumstances, the tangents remain intact: each point on the surface could be covered by a small piece of paper that is neither folded nor wadded into a ball.

However, in the process we look carefully at the folds of its projection. The stone Venus on the mantle appears to have folds when I look at it, yet its tangent planes are intact. The folds in the drawings of this book are artifacts of looking, or more precisely, they are artifacts of the drawing. Throughout the process the drawings are stylized, and there is some carica-ture. Even though the tangents are intact, the sphere, being infinitesimally thin, may pass through itself. The entire process is codified by means of the self-intersections and the folds. The fold set of the process forms a 2-dimensional surface (with seams). The double point set of the process forms a 2-dimensional surface. The triple point set is a single closed curve that intersects itself at a quadruple point. The interactions among these sets make the process interesting. The interactions among these sets contain the mathematical content of this work. I analyze these as invariant quantities of the sphere's evolution.

Chapter 4

Dimensions

The process of turning a sphere inside-out while keeping its tangent planes intact is called a *sphere eversion*. This process involves many dimensions, including a human dimension, which is the subject of the next section. This section is a playbill that describes the cast of characters, the actors who portray them, and previous work experience. These principle characters are the subject of the eversion, not the real people whom I mention in the sequel. By the end of the section, you may be dizzied in trying to keep track of each. In Chapter 8 there are synopses and reference tables. You may, during one of the intermissions, refer back here and refresh your memory. Also, the sets in question are drawn from several points of view and it is clear within the context of the pictures which set is which.

The sphere itself is 2-dimensional. The eversion is parameterized by time. At time $t = 0$, the sphere is a red sphere in 3-dimensional space. At time $t = 1$, it is a blue sphere inside 3-dimensional space. At each intermediate time, the surface of the sphere is bent, intersects itself, but its tangent planes remain *intact*: at each instant of time during the process, and for every point on the sphere, there is a neighborhood of the point for which the plane tangent to the sphere at the point approximates the image of the sphere; furthermore each tangent plane fits flatly into space as if it were the floor, ceiling or wall of a (tilted) room. The eversion represents a thickening of the infinitesimal sphere to a 3-dimensional object that has two spacial dimensions and one temporal dimension.

The red/blue sphere moves within a 3-dimensional space, and the time-elapsed version occurs in a 4-dimensional space-time. Despite popular accounts, not every 4-dimensional space is a space-time. For convenience, the process of eversion is considered to occur in such a space-time. But a mathematically sophisticated observer sees the process within a single hyper-solid

4-dimensional space. This section begins the descriptions therein.

You watch the sphere move. Each page of illustrations represents a photograph of that motion. These pages are the stills in an ordinary movie. Thus at each time, the sphere that sits in space between you and the page is projected to the plane of the page.

———————

The folds and cusps of some ordinary objects that are within my field of view such as a hand, a shadow, and a statue are described in Chapter 2. A sphere is an ordinary object, too. But its image during this deformation transcends an easy description. Rather, mathematical language, as the language of formulas and functions, gives precise descriptions that quantify the language of illustration. Here illustration will suffice since information is delivered from several different points of view.

———————

Consider the 3-dimensional space consisting of the images on the pages in the book. Each page describes the 2-dimensional projection of a 3-dimensional slice of a 4-dimensional process. The illustrations were prepared using a combination of cyan, magenta, yellow, and black pigments. The precise quantity of each type of ink in each illustration represents a point in a 4-dimensional color space projected onto the page. And yet again, the inks undergo a process as the pages are turned.

The projection of the picture to the page illustrates the structure by means of the sets of folds. These are line segments and circles illustrated by thick blue or red lines. As these folds pass into the interior of the sphere, they are depicted in ever-fading thicknesses and degrees of transparency — yet other dimensions. Remember the fold lines are the segments along which the tangent planes become perpendicular to your line of sight. As they evolve in time, they sweep out surface. The surface of the folds evolving has seams that are formed by the cusps evolving. The surface of folds is sewn together along the seams of cusps, and so the fold surface envelopes the evolution of projection.

At many stages of the process, the sphere intersects itself. The tangent planes at intersection points usually intersect as any two non-parallel planes would in space. Look at the corner of the wall in the room in which you are sitting. See the line segment of intersection between the south wall and the west wall. Imagine both walls extending beyond this corner. The intersection between the metaphorical planes is a line, and the planes themselves extend beyond your field of view. So the set of *double points* at any time form another 1-dimensional set. The collection of these as time moves forward forms another 2-dimensional surface. This surface is one of the most interesting ones encountered during the process. It is *non-orientable* or one-sided in the sense that there is a Möbius band within.

Also, the sphere may intersect itself as the south wall, east wall, and floor intersect at a single point. Such *triple points* stand isolated and therefore are 0-dimensional at any given time. But a point evolving in time forms an arc in space-time. So the triple point set of the process forms a 1-dimensional set. No triple points appear when the sphere is homogeneously red, nor when it is blue. So the triple point set, as a 1-dimensional set, forms a collection of circles in the interior of the process. In fact, it is only one circle, and at precisely one moment it intersects itself at a quadruple point of the eversion.

The quadruple point is the most critical of them all. At all times before this, the sphere can be pushed back to the red sphere without passing a quadruple point. Afterwards, the sphere must evolve to the blue side if it is to move away from the quadruple point.

The eversion presented here is quite a bit different from all of its predecessors in that its red side and blue side are not symmetric. In this sense, to complete this eversion, many more sketches were needed than in the predecessors. Even more strange is the fact that the blue profile does not appear for quite some time beyond the quadruple point. The process has some flexibility. Some critical areas can interchange their positions without damaging the process. But the manufacture of this eversion was quite labor intensive, and it took me and my collaborator, Sarah Gelsinger Brewer, quite some time to make sure every stage was complete. In fact, each drawing is a computation. One does not understand the complete computation until every individual calculation is complete, and one does not know which steps can be avoided until all the drawings are finished. An implicit invitation beckons: make the process as simple as possible while retaining the property of the singularities being isolated.

————

Each projection to the page can be sliced by a sequence of 2-dimensional planes. Each slice intersects the surface of the sphere in some simple closed curves (usually only one). These are circles in the plane that intersect themselves. The changes in circles give an alternative time direction. So the sphere eversion process can be thought of as a move to movies, and the dimensional metaphor changes from 3-spacial plus 1-temporal, to 2-spacial plus 2-temporal dimensions.

The distinction between spacial and temporal dimensions is a fictionalization. Since we are in the habit of thinking that the world is 3-dimensional and time is a 4th dimension, it is convenient to conceive of a process in 3 dimensions as a subset of 4-dimensional space. In the sphere eversion process, there is one preferred time direction: that which distinguishes red from blue. But at nearly each moment of that process, the sphere intersects itself, and that which faces us, occludes other facets of the surface. So to view the interior of the process, a height function is chosen in space, and a sequence of planes perpendicular to the height direction are chosen to intersect the sphere at that moment. The intersection of the sphere with these planes forms the self-intersecting circles. As the sphere changes from bottom to top, that processes is considered as being dynamic and therefore also temporal.

There are further illustrations. To indicate how each temporal slice of the process is mapped into 3-dimensional space, a set of instructions is drawn as abstract 1-dimensional sets on a sequence of spheres. Three separate 1-dimensional sets are given at each time: the blue and red fold sets and the pre-image of the double points.

In due time, each set — fold, cusp, double point, triple point, pre-image (known as the decker set) — and critical points thereupon are discussed in turn. For most of us, people are more interesting than sets, so let us first look upon the characters who made the story possible.

The Human Dimension

This is the story of sphere eversions since 1958. Nearly every person in this drama was contacted for commentary. The efforts to get at the truth, or at least a more interesting fictionalization peppered with personal stories, met with naught. Only a bit of minutia was uncovered in the correspondences. Nearly everyone involved is still living, and if I don't have the story exactly right, it is not because I didn't try. I am tempted to embellish the next few paragraphs just to get a reaction from the principals. Instead, I'll tell the story as it was told to me, as I have read the story on the internet, and as I remember it being told during various lunches and conferences. Any embellishments or half-truths are the responsibility of the lunch time stories, and I won't reveal those sources.

About fifty years ago, Steve Smale [Smale (1959)] proved the remarkable fact that the sphere could be turned inside-out without tearing or folding but allowing it to pass through itself. Raoul Bott was Smale's PhD advisor at the University of Michigan. It is told that Bott, upon hearing Smale's general result, examined its consequences in terms of the sphere and pronounced the result wrong because it was obvious to him that you can't turn the sphere inside-out without introducing a circle at which the tangent planes become singular. Smale persisted, convinced Bott, and earned his PhD. Smale went on to prove the high dimensional Poincaré conjecture and earn a Field's Medal. Since then many marvelous examples of sphere eversions in a variety of media have been given.

The story about Bott is also told about the referee of the original paper. It is rare that a paper's referee is the PhD advisor of the author. The story, then, has some apocryphal characteristics, so it entertains. "Your theorem can't be right! If it were, you could turn a sphere inside-out while preserving tangencies."

"My theorem is right, and *I* can turn the sphere inside-out."

Sphere eversions are thought to be impossible by the uninitiated since circle eversions are demonstrably impossible.

Some say Smale's proof of the eversion is only existential: His proof shows you that it can be done, not how to do it. Smale denies this. Smale's calculation is related to the way a certain loop in a space of transformations of 3-dimensional space can be contracted. One can trace through a variety of abstract spaces and functions between them and reconstruct an eversion. In particular, one of the more recent sphere eversions, that illustrated in Outside-In [Thurston et al. (1994)], uses the idea of "loop contractions" to give the computer animation. So maybe Smale's denial is vindicated.

The late Arnold Shapiro, who had a reputation of being particularly opaque with his explanations, described to Bernard Morin an idea to turn the sphere inside-out. Apparently, Morin did understand Shapiro's construction, because Morin and, subsequently, Marcel Froissart developed one of the more enduring examples of sphere eversions. Because of this construction, Bernard Morin was christened the "great geometric visualizer." Morin is the "great geometric visualizer" just as James Brown is the "godfather of soul." Bernard Morin happens to be blind.

Shapiro's construction involved an early 20th century idea that was due to Werner Boy who was the only student of David Hilbert to study geometry. Boy gave an example of a Möbius band whose boundary fit nicely on the surface of a sphere even though the Möbius band intersected itself thrice. The important feature of the Möbius band is that its tangencies are always (locally) embedded. Apparently, Shapiro's instructions were to paint the Möbius band as the intermediate stage. My own reputation for clarity (if I have one) is destroyed without the following detail.

A model of a Möbius band is often made from the margin of an ordinary piece of notebook paper. Cut the paper along the red line that defines the outer margin. Put a half-twist into the paper, and tape the ends together. In this way, the center of the Möbius band has circumference 8.5 inches. If by mistake two sheets of paper (temporarily stuck together) were cut, twisted, and taped together, then instead of a Möbius band a paper gasket that had a full twist would have been made. With two sheets of paper, there are a total of four edges that are taped together, and two pieces of tape are needed. The paper gasket is analogous to a gentleman's belt that has been buckled even though it has a full twist in it.

Mathematicians say that the Möbius band is covered in a 2-to-1 fashion by the twisted gasket. The two strips of paper, when taped, appear to be a Möbius band until they are unravelled. We also do not distinguish between a gasket or a cylinder; we call either *an annulus*. An annulus is the region that surrounds a (1-dimensional) circle in the plane. The circle can be represented by the equation $x^2 + y^2 = 1$ in the plane, and an annulus is represented by the inequalities $\frac{1}{2} \le x^2 + y^2 \le 2$. A model of the annulus can also be obtained by cutting a hole into a paper plate, or by removing the bottom of a paper cup. The tropical region of the earth represents an annulus.

Boy's surface allows the Möbius band to intersect itself in a peculiar, but nevertheless a nice way, and the doubly covering annulus can be carried through that construction. (To construct Boy's surface at this point in the discussion would take us far afield. I have written about it elsewhere [Carter (1995)].) The boundary of the Möbius band is a single circle that, under Boy's construction, lies on a sphere in space and so bounds a disk on that sphere. The boundary of the double covering annulus consists of a pair of circles. These lie immediately above and below the sphere, and each bounds a disk — one in the region above the sphere and one in the region interior to the sphere. The circle boundary of the Möbius band that lies on the sphere resembles, to some extent, one of the two leather pieces that forms the surface of a baseball.

When two disks are attached to an annulus, a sphere is formed. Consider a paper cup. The bottom of the cup represents one of the disks attached, and a lid for the cup represents the other. Alternatively, consider the earth made of a tropical region (that is an annulus) and polar disks (allow me some latitude to match the latitudes). The annulus that is mapped into space as a double covering of the Möbius band via Boy's construction can be capped off by two disks above and below the baseball seam on the sphere.

Now, remember that all the mathematical surfaces are infinitesimally thin. The sphere that is formed from the doubly covering annulus can pass through the intermediate Boy's surface which is "one-sided." The sphere passes through the Boy's surface in a moment of serendipity and simultaneity. In doing so, it has turned inside-out.

The unfortunate aspect of the preceding paragraphs is that it describes the easy stage of Shapiro's idea. Moving a blue or red sphere into the position at which it double covers Boy's surface is much more difficult. I, personally, don't know how to do this because there are eight triple points that form the corners of a cube. I just am not sure how to get an embedded sphere to that position.

Tony Phillips heard about Shapiro's construction and he independently used Boy's surface in his Scientific American article [Phillips (1966)] on the sphere eversion. Phillips's article was very influential on a number of young people, myself included, who were learning that there was a new type of math out there called topology. It involved stretching and bending without tearing and it required using techniques that until modern times had been unknown. To a child who grew up learning about the great new world explorers, who was fascinated with the lunar project, and who was looking for new frontiers, this mathematical world was full of the promise of excitement. It still is. Phillips's article did not use what we now call the movie moves, but it did illustrate each stage of the eversion by using a sequence of cross-sections. One commentator says that it is not particularly easy to see how to get from one stage to another. Someone whom I know says that the illustrations in the Scientific American article contains known mistakes.

The current eversion explicates each step in the process by using a finite set of moves. But it is written in the spirit of Tony Phillips's article.

I have been told that Morin calls the eversion (that is closely associated with him) the Froissart-Morin eversion. My source said that Morin practically insisted upon this name. It may be an instance of Arnold's principle which states that if a mathematical concept is named for a particular person, then the actual idea was first discovered by someone else. Arnold's principle is said to apply to itself!

Charles Pugh, then a graduate student at Berkeley, made a collection of chicken-wire models that were hung in a hallway at the Berkeley Mathematics Department. I recently met someone who remembers them being there one day and gone the next. These were models of the Froissart-Morin eversion which depends upon a symmetry between the red side and the blue side. Once at the half-way stage, the model can be rotated in space, and the process reversed. Nelson Max carefully measured each model and recorded the data. He and Tom Banchoff produced an animation based upon these measurements. This early film was an amazing piece of computer animation/computer art. I don't think there would be a *Toy Story* without this pioneering work.

That the chicken wire models were stolen and lost is a real pity. In a Circle of Hell, there is a thumbless thief who must reconstruct them from memory with a pair of blunt wire cutters. The thief completes them and each night forgets the sequence. His or her hands are punctured and torn by metal scratches — these wounds do not heal.

George Francis is a fellow who has made much about sphere eversions. He realized that Shapiro's description and the Froissart-Morin eversion fit into an infinite family of eversions that he called the *tobacco pouch* eversions. In addition, Francis was one of several people (*e.g.* Rob Kusner and John Sullivan) who used an energy functional to deform Morin's half-way model to the red embedding and the blue embedding. This eversion is called the *Minimax eversion*.

An additional scholar of sphere eversions is François Apéry. Apéry produced, among other amazing things, a piece-wise linear model for a sphere eversion. The meaning here is that the sphere is approximated in a box-like fashion, and each stage is created by functions that not only have tangencies, but are often modeled by linear functions.

One of the most popular of the sphere eversion animations is the Outside-in video. The idea for this sphere eversion is attributed to Bill Thurston. One of his sons, Nathaniel Thurston, wrote a substantial amount of the computer code for Outside-in. The video was produced at the Geometry center with the help of a number of other people: David Ben-Zvi, Silvio Levy, Dell Maxwell, and Tamara Munzner are the names most frequently mentioned, and a full set of credits can be found at the end of the film which is available on-line via a Google search on Outside-In.

A purple and gold sphere is turned inside-out through a series of crenelations or corrugations and a parameterization of the belt trick. The belt trick is fairly standard example in mathematics, and it should be familiar to anyone who owns a garden hose or a central vacuum in their home. If your garden hose has a kink in it, you can undo the kink by twisting the tip of the hose a full rotation in the opposite direction of the kink. The fact that you can hold a glass of water in the palm of your hand, and make your arm twist two full cycles without spilling the water is another manifestation of the belt trick. Its relevance to the sphere eversion can be seen in two ways. The most natural way of turning a sphere inside-out would be to put a full kink in the equator. Bill Thurston, when Nathaniel was between 8 and 13 years old, had described an idea for a sphere eversion that used the belt trick to uniformly smooth the kink out of the equator. The other way that the belt trick comes into the sphere eversion is directly through Smale's original work. One can trace through a complicated argument to see that Smale's computation depended upon this loop of symmetries of space being contractible in the space of all symmetries. Bill Thurston exploited this to give a concrete description of sphere eversions in terms of the belt trick. I saw this first mentioned in John Hughes's 1982 dissertation. John says that he learned it from Bill, and Nathaniel's younger brother Dylan told me that Nathaniel was 13 years old in 1982.

———————

Both the Minimax eversion and Outside-In result from work initiated at a place called the Geometry Center, a location in Minnesota that was funded through the National Science Foundation. The Geometry Center was a fun place to visit: a toy room for geometrically-inclined mathematicians. Many geometric objects could be manipulated and viewed on a computer screen. Some of the most popular and enduring mathematical imagery from that era was produced at the center. My opinion is that the toys were provided, the mathematicians were just getting the hang of them, and the Center shut down before anyone could envision the potential of the machinery. There are many aspects of mathematical research. One, and the most important of these, is the ability to produce theorems — results that transcend specific examples and that are timeless. Another is to produce enduring examples. Examples are the testing grounds for theorems. A deep understanding of intricate examples can lead to more theorems as well as make the meaning of the theorem more accessible. Genuine scientific research is guided by intellectual curiosity, does not follow a time clock, and may be best achieved when a scientist has the intellectual freedom to explore without encumbrances.

The essential steps of the eversion that is presented here first appeared in my book, "How Surfaces Intersect in Space," [Carter (1995)]. That book was written shortly after the first "movie move" theorem had been proven. A number of parts of that eversion did not satisfy me. Most importantly, one step in particular skipped essential details. After the second of the "movie move" theorems had been proven, I realized there was a much more rigid method of everting the sphere. One which took the fold set and the cusp set of the projections into consideration would be superior in a number of ways. Most importantly, the complete internal structure could be analyzed.

The two "movie move" theorems were joint work, the first [Carter and Saito (1993)] with Masahico Saito, and the second [Carter, Rieger, and Saito (1997)] with Masahico Saito and Joachim Rieger. In the first movie move theorem only a sense of height was analyzed in relation to moving a surface around. In the second movie move theorem, we also fixed a sense of height in the cross-sectional stills. Since the sphere eversion is such a complicated process, it is a good testing ground for the movie move theorems. Both versions of the movie moves theorem were developed in order to (a) develop a calculus for moving surfaces that are embedded or immersed in 4-dimensional space, (b) to recognize when two surface diagrams represent the same embeddings (the diagrams are 3-dimensional pictures or sculptures that represent surfaces embedded in 4-space), and (c) to develop an algebraic method for distinguishing such surfaces. As it turns out, a number of other mathematicians were interested in these processes for their algebraic and categorical ramifications. So what started out to be an excursion in order to better understand surfaces as they are embedded in 4-dimensional space turned out to be a fundamental axiomatic system for so-called higher dimensional algebra. The sphere eversion is a calculation in that higher dimensional algebraic system.

To conclude this section and turn to the next topic, the current book is an intricate, intimate, and taxonomic view of an example of the sphere eversion. The example is based on, but not identical to the Froissart-Morin eversion. Successive stages differ by as little as possible. Usually only one step is performed at a time. Sometimes, a few steps are performed simultaneously when the sequence of these is obvious from one of the several points of view given. In the next section, the possible steps that can be taken and the types of images seen are given. The imagery used in describing these steps is literally a collection of images, but it also involves the use of language. Some of the language is standard mathematical terminology, and this terminology needs supplementing. Let us proceed.

Chapter 5

Immersed Surfaces

Surfaces are approximated by their planes of tangency. In order to understand how a surface can intersect itself, it is necessary to understand how planes can intersect. The intimate details of the surface intersections are encapsulated in the intersections between two planes and among three planes. The large scale structure may undulate or curve back upon itself, but in a local coordinate system it appears that planes intersect.

It is possible for a pair of flat planes to lie tangent, but when that happens the planes coincide. When the book is closed, the pages are apparently tangent planes, but the pages have thickness. If the pages were infinitesimally thin, then they would be coincident and indistinguishable. Coincident pairs of planes and parallel planes are the exception. In a truly technical sense, two planes chosen at random intersect along a line. The angle of this intersection is not an issue here. The model to consider is the intersection between the south wall and the west wall. The line of intersection is vertical.

In coordinates, the south wall is represented by the plane $x = 0$, while the west wall is represented by the plane $y = 0$. The set of points that lie along the intersection of these two planes is the line $\{(0, 0, z)\}$ where z can be any real number. An illustration is indicated in Fig. 5.1.

Similarly, three planes *can* intersect along a line. This would be the case if you were to put a diagonal Japanese screen up in the southwest corner of your room, and arrange it so that its edge lies along the corner of the walls. More often, three planes that intersect do so at a single point in a way similar to the intersection of the floor, west wall, and south wall. In the same coordinate system as above, the floor has equation $z = 0$, and the intersection point among these three planes is $(0, 0, 0)$ — the origin of space. An illustration is indicated in Fig. 5.2.

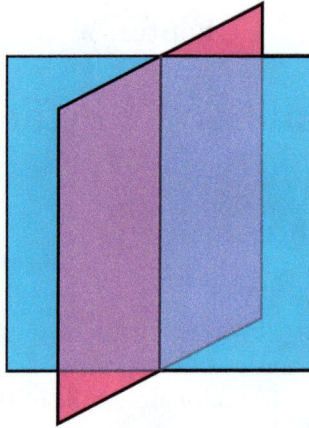

Fig. 5.1   An arc of double points

The two illustrations depict the local pictures for a self-intersecting sur-face in the case that the tangencies of the surface remain intact. They illustrate *double points* and a *triple point*, respectively. As the sphere turns from red to blue, it intersects itself in local pictures in one of these ways. At each time, the set of double points is one dimensional, and at each time, the triple points are isolated even though they also lie along arcs of double points. As the double points evolve in time, they form a surface, and as the triple points evolve in time they form a 1-dimensional set.

———

It is also good to demonstrate the things that cannot happen when tangencies are preserved. Figure 5.3 indicates a type of point called a *branch point*. At this point a segment of double points ends at a peculiar singular point. At precisely this point of the surface, the tangent plane is not to be found. In Chapter 9, I show you how not to evert a sphere, but how to turn a sphere inside-out, by allowing a pair of branch points to appear in an intermediate stage. Observe that the fold line at the branch

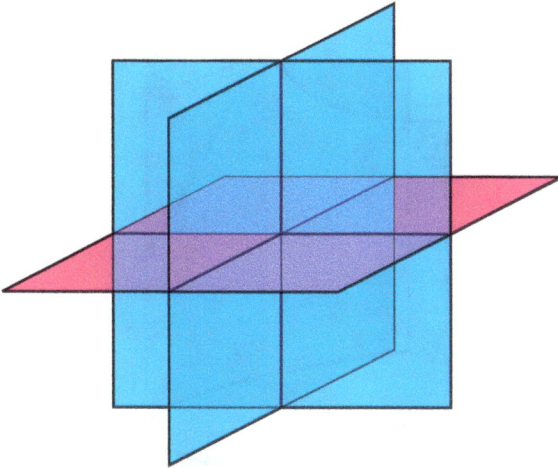

Fig. 5.2 A triple point as the intersection of coordinate planes

point changes from red to blue. This change in color indicates a singularity in the tangent direction. At a branch point, then, the tangent plane is not uniquely determined; a sphere turned inside out via the introduction of a pair of branch points, does not have its tangent planes intact at the branch points. It is not an eversion.

———

A sphere that intersects itself in space in such a way that it has only double points and triple points of the type described above is said to be a *general position immersed sphere*. During the process of the eversion, the illustrations given are general position immersed spheres. At the points in which the illustrations change (the part of the process that is not illustrated, or the part of the process at which your brain is meant to interpolate), the sphere is immersed, but it is not in general position. At these stages, there may be simultaneous tangencies between pieces of the surface, or there may be a quadruple point.

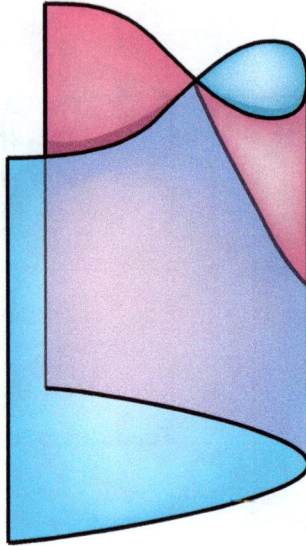

Fig. 5.3   A branch point: this does **not** occur during the eversion

A surface when projected to a plane, like the orange that we look upon, may have a segment upon which the tangent plane is perpendicular to the plane of projection. Such a segment is called a *fold line*; an example is illustrated in Fig. 5.4. The important fact to understand is that a drawn surface, or a surface that is seen, has a fold. This fold is the profile that distinguishes the surface from its surroundings.

In addition to metaphorical language, formulas can be used to describe the ideas of a fold. The expression $y = x^2$ represents a parabolic cylinder with its axis lying along the z-axis in space. When this surface is projected onto the yz-plane — the plane for which $x = 0$ — two sheets overlap when $y > 0$; meanwhile, along the z-axis ($x = y = 0$) the surface folds onto a line. This surface is the local model of a fold just as the intersection of the pair of planes $x = 0$, and $y = 0$ is a model for a double point, and the intersection among the three planes $x = 0$, $y = 0$, and $z = 0$ is a model for a triple point. At any particular time, a fold is a 1-dimensional set; as the sphere moves, the folds form a surface in space-time.

Fig. 5.4   A blue fold

———————

Two folds converge at a *cusp*. One fold is visible, and that fold is colored by the visible side of the surface. The other fold is obscured in the projection and has the opposite color. The cusp represents a point of the projection of the sphere onto the page at which the tangent plane vanishes in the projection. The tangents in space do not vanish, but they do vanish in the projection. Cusps can be given formulaically: Consider the expression $y = x^3 - zx$ over the set of points with both x and z taking values between -1 and 1. When this surface is projected into the (yz)-plane (for which $x = 0$) in space, there is a cusp at the origin. Figure 5.5 illustrates. In this figure the cusp has been drawn so that the blue fold line is visible. The fold is blue in the sense that it appears on the blue side of the surface. The fold that is not visible, from the point of view of the observer, is red. The *color of a cusp* is the color of the fold that is visible. Cusps are isolated among themselves, but are the points at which two folds converge. At a particular time, the cusp set is a collection of isolated points, and as the sphere moves in time, the cusps form a 1-dimensional set.

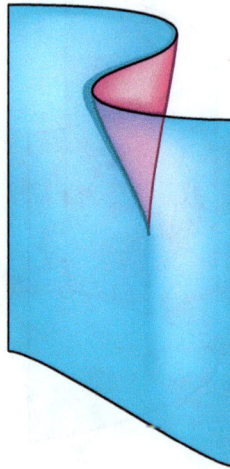

Fig. 5.5   A blue cusp

Figure 5.6 illustrates an immersed sphere that is one of the steps of the eversion. The figure demonstrates a red fold, a red cusp, a double point arc, and a triple point by enclosing these in green squares. The triple point indicated lies below a red sheet. When a point of interest lies below a sheet of the sphere, it is said to be *veiled*. There may be several veils between you, the viewer, and the interesting point.

There are three other triple points in the figure; try and find them. The cusp at the bottom of the figure is blue, and the red sheet veils it. On the right of the illustration a blue fold emerges from behind a red fold and this blue sheet intersects a red sheet towards the bottom of the illustration. This particular phase in the process is among the more complicated of the illustrations. It, and each of its predecessors and descendants, are also described via a sequence of horizontal cross-sectional planes that intersect the surface in a sequence of self-intersecting closed curves. The constituent pieces of such intersections are described momentarily.

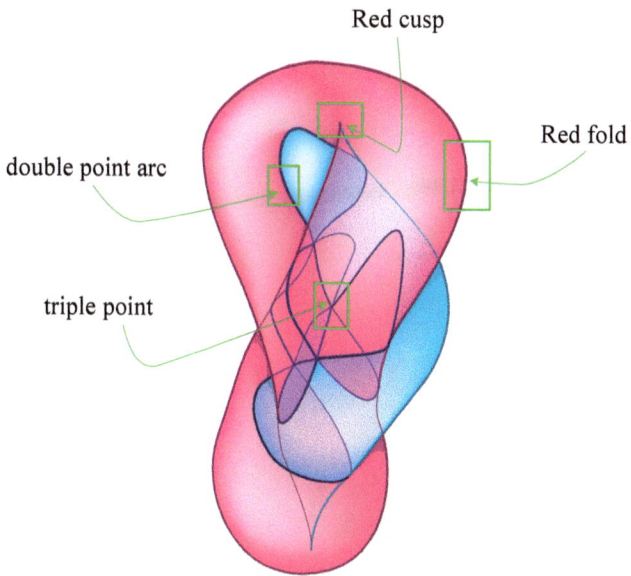

Fig. 5.6   Cusps, folds, double points, and triple points together

Chapter 6

Movies

A surface that is immersed in 3-dimensional space may have double points and triple points. To be *immersed* each point on the surface has a neighborhood so that the surface is approximated by its tangent planes within such a neighborhood, and each tangent plane is embedded in space as any flat 2-dimensional subset might be. If the only points of self-intersection are double points and triple points (as described above) the surface is *in general position*.

During the eversion, the red sphere moves through immersions until it becomes blue, and each illustration that is given here is the projection (onto the page) of a general position immersion. Two successive pictures differ by an immersion that may not be in general position. So the set of illustrations represents a *movie* of the sphere eversion. The interesting parts of the movie go unseen. You, as the observer, learn to interpolate between stills of the movie.

Each still is the 2-dimensional projection of an immersion of a sphere. To see the projection without seeing the internal structure is to miss the nuance of the actors' interpretations. For example, Fig. 6.1 indicates the same surface that Fig. 5.6 does, and a different triple point is indicated since the one that was behind the veil now is completely hidden: the first layer of surface is opaque.

Three complementary techniques illustrate the internal structure of each immersion. First, the surfaces themselves are illustrated as if they are transparent. Second, the surface information is removed, and the fold lines, double points, and triple points are indicated. Fold lines are colored with the color of the surface that is on the convex side of the fold. Degrees of transparency and line thickness convey the distance of these 1-dimensional sets from you, the observer. The same fold lines, double curves, and triple

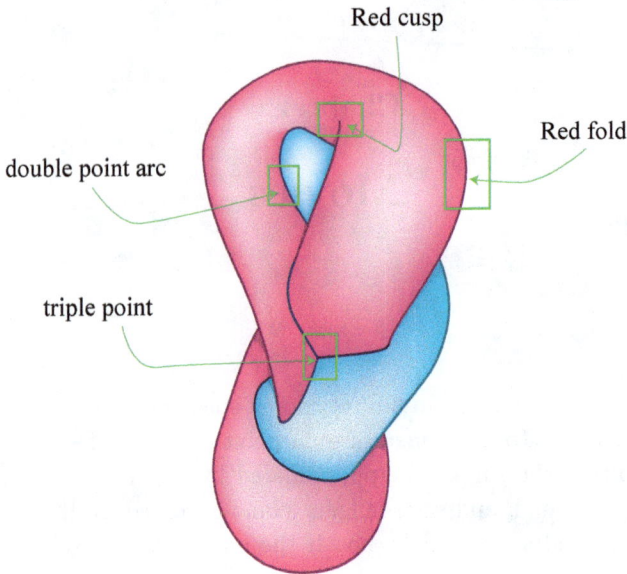

Fig. 6.1   An immersed sphere with its inner structure hidden

points are indicated in the semi-transparent illustrations, and these sets with their color and transparencies given are almost enough to reconstruct the internal structure. Sometimes though, some critical levels on the fold set are difficult to quantify. Third, the immersed surface is considered to lie in front of the page, and a sequence of horizontal planes that lie perpendicular to the plane of the page slice the surface in a sequence of immersed circles. The horizontal planes are chosen to lie above and below each critical level of the immersed surface. And the sequence of immersed circles is called a *movie of the surface.* To fully describe the notion of a movie, the notion of critical level is given a precise definition, and a taxonomy of the types of critical points is necessary.

———

In ordinary usage, *a critical point* means a time or a place at which a decision is made or a change is happening. The ordinary usage parallels the technical usage in which a critical point is a point at which a function changes directions or a point at which a derivative vanishes. Each immersed sphere is projected to the plane of the page, and the page has a notion of *height* defined upon it. This way: ↑ is up. The *critical points of the height function* are the points at which some sets change from going up to going down or vice versa. The sets in mind are the fold set and the double point set. A triple point is also viewed to be critical since three sets converge there.

The height functions on the homogeneously red sphere and the homogeneously blue sphere each have exactly two critical points: A maximal point and a minimal point. Reading from bottom to top, the minimal point is the *birth* of a simple closed curve, and the maximal point is the *death* of a simple closed curve. The cross-sections and the surfaces are indicated for the red sphere in Fig. 6.2.

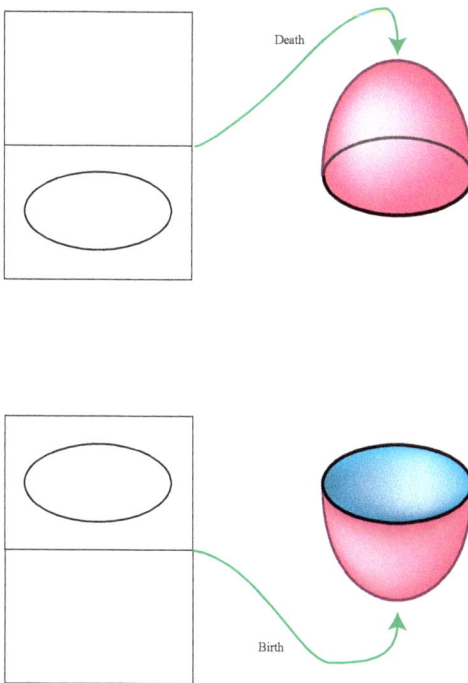

Fig. 6.2   Birth and death critical points

A *saddle point* occurs at the crotch of a pair of pants, the arm pit of a shirt sleeve, or indeed on a horse's saddle. The English saddle, which is less ornate than its western counterpart, has a pommel at its front and a cantle at its rear, and the seat curves downward to the left or right and upward towards the cantle and the pommel.

As the sphere moves it develops saddle points which persist for some time during the motion, but eventually these disappear as the blue side dominates. In a movie of a saddle, a pair of fundamentally horizontal arcs are replaced by a pair of arcs that are for the most part vertical, or vice versa. At such a saddle point, the fold set has a critical point.

These critical interactions require some more description. A *still* of a movie of a sphere is a planar picture of a closed curve. There is a *direction* that is defined on a still: →, left to right. Arcs on the closed curve are "usually" moving left to right or right to left. The points at which directions change are critical points for the direction function. Within the stills, vertical lines are disallowed since their tangencies would be singular for the direction function. Instead, the curves turn at critical points. There is no intrinsic prejudice about vertical versus horizontal; arcs that are purely horizontal are also disallowed. But who makes the rules and how are they made?

An immersed closed curve in the plane is defined by means of a function that is supposed to have non-vanishing tangencies at each of its points. Such an arc is approximated by its tangencies just as the sphere is approximated by its tangent planes. With respect to any preferred direction, the tangent direction, then, is almost always pointing towards or against the preferred direction. The tangent direction is usually not "strictly parallel" to the preferred direction, nor is it strictly perpendicular to the preferred direction. This non-parallel property is a quasi-physical phenomenon. Place a piece of paper on the floor beside you and drop toothpicks upon the paper. The toothpicks usually neither land vertical nor horizontal with respect to the orientation of the paper. If the toothpick does appear vertical, then a careful measurement indicates that it is only vertical within some small measurement error. Mathematically, we say that a line chosen at random is probably neither horizontal nor vertical. Let me exemplify the meaning of the word *random* in this context: the slope of a line is a random real number, and if the slope appeared to be 0, then each of its infinitely many decimal places would have to be 0. The probability of choosing digits at random and having them all be 0 is infinitesimally small.

Now as the curves are being drawn, I often *choose* to draw the tangents as either horizontal lines or vertical lines because by doing so I can control the shape of the curve. The curves are computer generated images, and their tangent directions are not chosen to be arbitrary real numbers but they are chosen from a finite (albeit large) set of rational numbers. The points for which I choose to have vertical tangents are always among the critical points of the direction function. The drawing software interpolates curves between chosen points dependent upon a choice of tangent direction, magnitude, and rate at which the tangent changes. If the first point has a tangency pointing left and the second is drawn to the right and has a tangency pointing right, then the computer interpolates a farthest left-most point between them. Such a point is a critical point for the direction function.

So when a pair of arcs that, for the most part, have horizontal tangents are replaced by a pair of arcs that each have a point of vertical tangency, there are a pair of critical points that are born. The birth of these critical points on the interpolating surface forms a critical point on a fold line.

When I draw the stills to a movie, I am assuming that you, the observer, are viewing the still from the bottom of the rectangle that contains the still. The arcs are cross-sections to the sphere pigmented on its inside and its outside. So the closed curves should be similarly pigmented, but they are not. Instead, the critical points of the closed curves are tacitly pigmented via the corresponding folds, and there is a simple convention that determines the pigments of the fold: The color of a fold, or the color of the critical points in the cross-sectional stills in a movie representation are determined to be the colors on the *optimal side* of the critical point. If the arc in the still that faces you is red (respectively, blue), and the critical point on that arc is a minimal point, left-pointing, or ⊂, then the critical point is red (respectively, blue). The same conventions apply to right-pointing: ⊃ maxima.

Figure 6.3 depicts a blue saddle as described above. I ask you to imagine red saddles and saddles that are obtained from this movie by reversing its time direction.

—————

There are eight different types of cusps that can occur. Figure 6.4 indicates a minimal cusp in which the blue fold is visible and on the left. The other seven variations occur when (a) red and blue are interchanged,

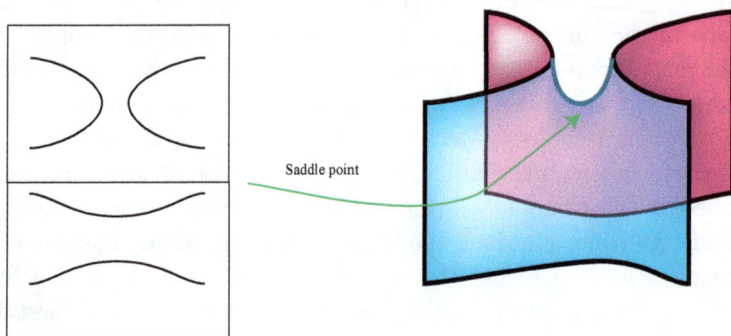

Fig. 6.3   A saddle critical point

(b) the visible fold is on the right: $\supset$, (c) the cusp is a maximal point and the visible fold is on the left: $\subset$, or (d) the cusp is a maximal point and the visible fold is on the right: $\supset$. You are encouraged to draw these variations of Fig. 6.4.

The *color of a cusp* is the color of the more visible fold that emanates from the cusp. Thus the cusp illustrated is a *left-down-blue cusp*.

———

Figure 6.5 indicates the standard illustration of a (blue) *torus* which is the surface of a (blueberry) doughnut or bagel. The illustration is decomposed in terms of its visible and invisible folds, and a movie presentation is indicated. The critical points are labeled with numbers 1 through 10, and each represents a change between the stills. In order (from bottom to top) these are:

(1) the birth of a simple closed curve
(2) a down-pointing cusp with a visible ($\supset$)-red fold once veiled by a blue surface
(3) a down-pointing cusp with a visible ($\subset$)-red fold once veiled by a blue surface
(4) a fission saddle point (one curve becomes two)
(5) the interchange of two ($\supset$)-folds on the left of the figure
(6) the interchange of two ($\subset$)-folds on the right

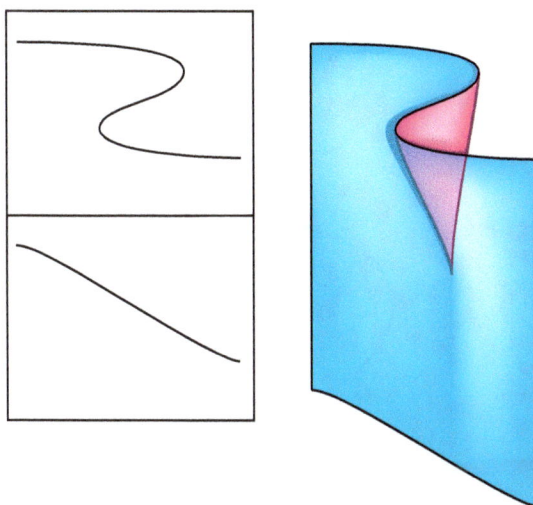

Fig. 6.4 A blue minimal cusp with visible fold on the left

 (7) a fusion saddle (two curves merge to one)
 (8) an up-pointing cusp with a visible ($\supset$)-blue fold on the right
 (9) an up-pointing cusp with a visible ($\subset$)-blue fold on the left
(10) the death of a simple closed curve.

Although the torus is not a central figure in the discussion of the sphere eversion, its depiction is a standard example throughout mathematics classrooms and topological lectures. My hope is that the current short analysis — when there are only few details upon which to focus — helps you understand the context and conventions of the subsequent drawings.

Double Points and Triple Points

An immersed surface in 3-dimensional space has a closed 1-dimensional set of double points. That is, the double points form circles that may have further self-intersections at triple points, and that may have many critical points, but always form closed curves. Such curves have maximal and minimal points which, from the point of view of a movie, are caused by a

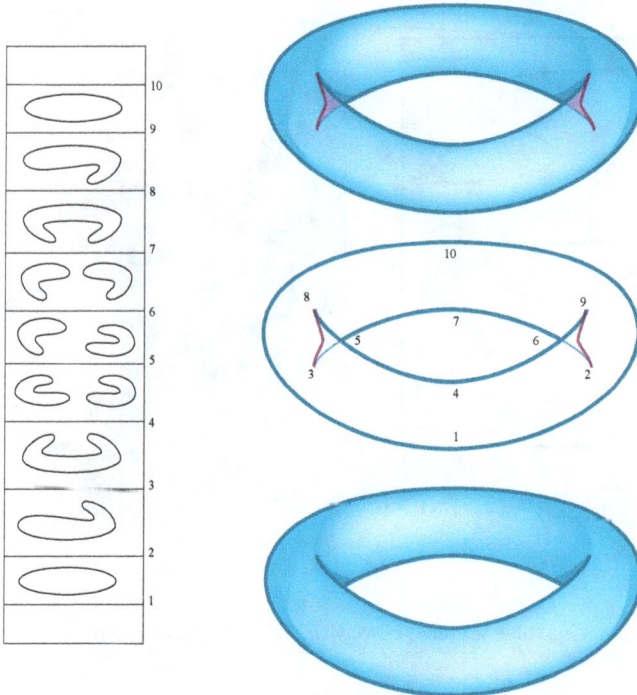

Fig. 6.5   The standard torus and its corresponding movie

pair of parallel strands crossing back and forth or a pair adjacent crossings disappearing. Figure 6.6 indicates the birth of a pair of double points (from bottom to top) in the case that the two sheets involved have different colors. As an exercise, the variations of this scenario are left to you. The illustration is called a *type II birth*. A *type II death* is the upside-down version of the movie.

Figure 6.7 indicates a triple point in movie form. The three sheets that intersect are no longer flat coordinate planes, but each sheet is bent. Nevertheless, the triple point appears to have three flat sheets coinciding

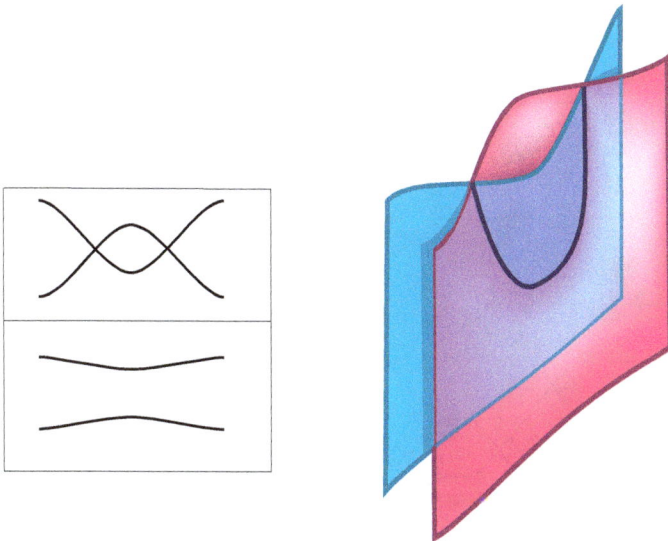

Fig. 6.6   The birth of a pair of double points

at the critical level of the picture. The illustration is called a *type III move*. Observe that the three sheets involved could all be the same color, or any two of them could be colored the same.

————————

Double point arcs can cross folds. Figure 6.8 indicates this situation which is called a ψ-move, a *double point bounce* or simply a *bounce*. Among these names, the name ψ-move is the most precise: The critical level from one perspective resembles the letter ψ. However in the projection, the fold lines and the double points appear to become tangent. So the two sets appear to touch and bounce off each other. As much as I would prefer to use the more technical term, I find myself thinking, "the critical point bounces inside." Words and terms sometimes acquire power beyond their chosen context, and it becomes difficult to disassociate the word from the idea. I know full well that the double point moves through the critical point, but my inner voice always calls this a bounce. So the terminology, while not fully descriptive, stays.

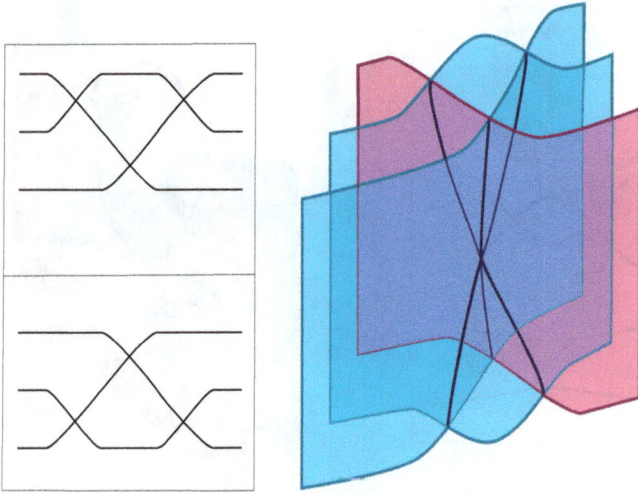

Fig. 6.7   A triple point

Critical Exchanges

In our own universe, the order in which some things are done does not appear to matter. If, when you set the table, you first put out the spoons and then the forks, the result is the same in the case when you first put out the forks. On the other hand, it is a good idea to check if the keys are in your hand before you lock the door. Some critical levels on a given immersed curve may be interchangable: a pair of minima ($\subset$ and $\subsetneq$) occurring sufficiently far apart in a vertical direction may be interchanged. Similar interchanges occur for a pair of distant maxima ($\supset$ and $\supset$), a distant pair of a maximum and a minimum ($\supset$ and $\subset$), and an optimum and a crossing that are far enough apart. How far is far enough? If the immersed curve could be drawn with these events occurring at the same point from left to right, then there are two perturbations of that scenario. The critical exchange is the manifestation of these perturbations. Figure 6.9 illustrates three such cases. There is a fourth case, and there is the obvious variations of the exchange of a crossing and a maximum in which a crossing and a minimum exchange positions.

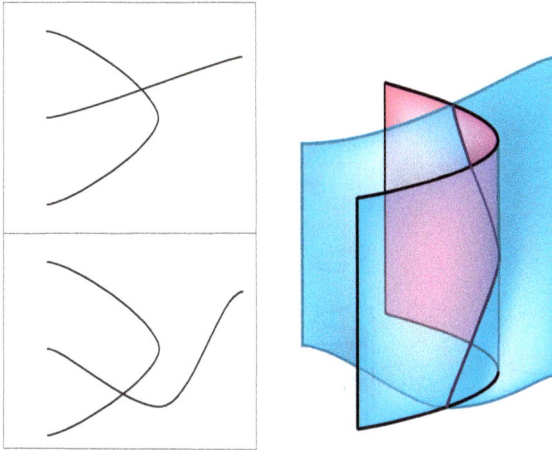

Fig. 6.8 A ψ-move in which a double point segment crosses a fold

Example

Figure 6.10 contains an illustration of a movie and the corresponding immersed sphere. The critical levels of the movie are numbered and the corresponding critical points on the sphere are illustrated. There are several places at which critical exchanges are not explicitly listed. Also the critical levels (from left to right) within some stills of the movie are out of order. Finally, the order of events within the movie may not strictly coincide with the height on the corresponding illustrations. The orders of the critical points can be exchanged when they are far enough apart on the surface. I have some good reasons for these inconsistencies of notation, and I have some bad reasons. You can decide which is which.

Sometimes I get lazy. More specifically, the precise placements up and down of the critical levels are determined by editing a previous figure. In the editing process, critical points are moved. The lines that connect the critical points are known as "Bézier" curves. They are controlled by various mouse control clicks and visual sense. To put it plainly, I choose the positions of the critical levels on the surfaces to make them look as good as I can

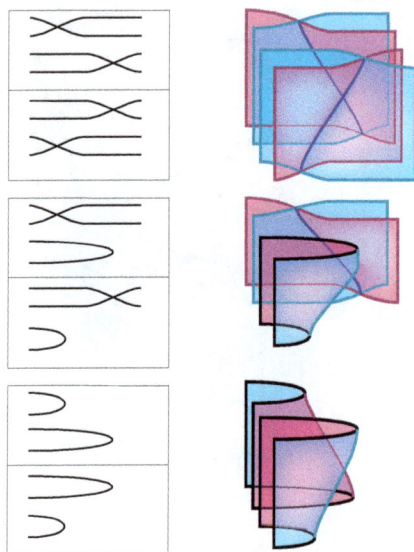

Fig. 6.9 Several critical exchanges

while using Bézier curves, and while trying to keep the number of pieces that must change to be as few as necessary. Similarly, the curves in the movies are chosen to look nice, and to be topologically accurate. But again, movies and the surfaces are edited from a previous picture. So steps are skipped. The figures don't look as nice if a conventional set of drawing tropes is adhered to religiously. Also, the order of critical levels may have come from a previous figure and the levels in the movies may have come from a subsequent figure.

The movies themselves would become painfully long if each critical exchange is explicitly listed. On the other hand, many changes between surfaces (movie moves) can occur when various events exchange positions. These extra exchanges would result in way too many surfaces being drawn. While the beginning algebra student may enjoy seeing each step in a calculation being worked explicitly, it does not take long for one to see that steps can be safely skipped. Virtually all the skipped steps herein have to do with critical exchanges.

My own visual sense and kinematic imagination is fairly well tuned. These visual abilities come from practice. There is a didactic purpose to the current work. I am training you to see the relationships between the movies and the illustrations. The missing steps and inconsistent height functions help you know what features to find, and help train your visual mind. The current work is the most detailed and complete version of the sphere eversion that has been created. It could be improved, but this is as far as I go with it.

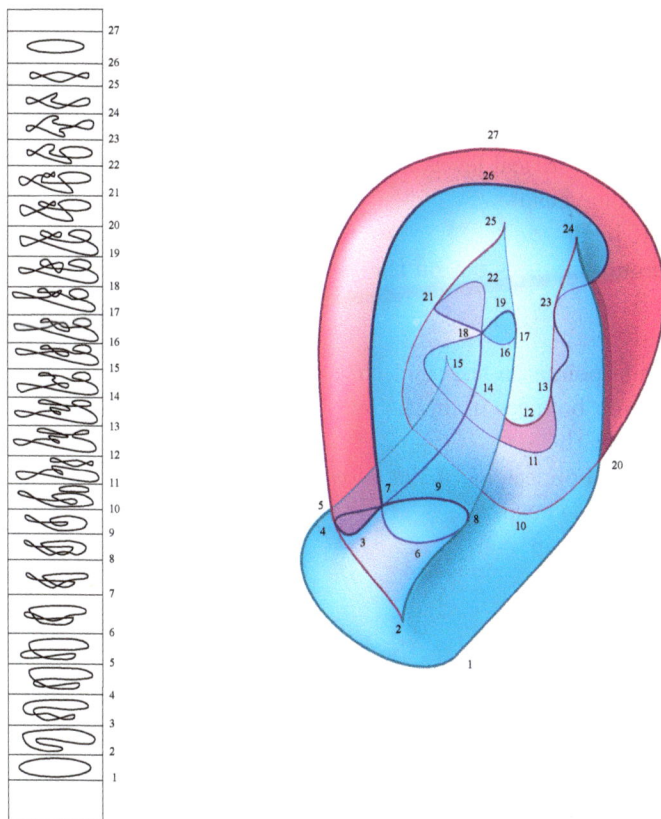

Fig. 6.10 An example of a movie and the corresponding immersed sphere

Here is a list of the critical levels:

(1) birth
(2) right-down-blue cusp

(3) type II birth
(4) bounce on the left
(5) critical exchange
(6) type II birth
(7) type III
(8) bounce on the right
(9) type II death
(10) birth
(11) type II birth
(12) saddle point
(13) minimum bounces out on the right
(14) interchange critical levels in the still
(15) left-up-blue cusp
(16) type II birth
(17) Maximum bounces out on the left
(18) type III
(19) type II death
(20) maximal points on the right change position
(21) mimimum bounces in on the left
(22) type II death
(23) minimum bounces in on the left
(24) right-up-blue cusp
(25) left-up-red cusp
(26) type II death
(27) death.

Summary

A general position immersed sphere in 3-dimensional space potentially has double points and triple points. By looking at a sphere, we are projecting it to the plane of the retina. In this projection there are fold lines that either form closed curves or that end at cusps. A sequence of cross-sectional planes containing the line of sight intersect the sphere in immersed planar curves. These have minima that point left, $\subset$, and maxima that point right, $\supset$. Successive stills differ by (1) births, (2) deaths, (3) saddles, (4) cusps, (5) type II moves in which a pair of double points appears or vanishes, (6) type III moves in which three points that form a triangle interchange their relative position and the triangle is reflected, (7) points at which double points pass over folds, or (8) critical points within the stills can change relative position. These changes in cross-sectional views are reflected in critical points on the projection. For such a sphere the cusps and the triple points are isolated points. The double points form closed curves. The folds are closed curves and arcs that terminate at cusps. The movies are read from bottom to top, and the viewer is assumed to read the cross-sectional still from the point of view of bottom of each still. Depth within the immersed sphere is indicated by making lines appear thinner and more transparent.

To help you understand the evolution of the immersed spheres, I describe the changes that occur between the pictures and their corresponding changes to the movies. These movie moves are the subject of the next chapter.

Chapter 7

Movie Moves

Each immersed sphere that is viewed may have double point arcs and isolated triple points. The sphere's projection along your line of sight has isolated cusps at which pairs of fold arcs converge. This chapter describes how these singularities are introduced and how they change as the sphere evolves from red to blue.

Thus, I describe three types of interactions among the singularities: interactions in which only the fold and cusp set is affected, interactions in which only the double points and triple point sets are affected, and interactions in which the folds, cusps, double points and triple points are affected. Each change to the singular sets is named, and in Chapter 12 each step in the process is described by means of one of the interactions given here.

The first section gives a digression on how the folds, double points, and triple points are drawn on pieces of the intrinsic sphere. As the sets evolve, an interpolating (or time-elapsed) view of these sets is demonstrated. These time-elapsed viewpoints help develop a temporal-to-spacial intuition. As time passes in the process, the "camera shutter" is left open, and the change in the interesting sets is depicted from left to right. In these views, the cusps, folds, double points, and triple points of the before/after scenarios form the boundaries of a set of one larger dimension.

The Evolution in the Intrinsic Sphere

To say that the sphere is 2-dimensional is to say that it can be decomposed as a set of patches, each of which is identifiable with a portion of the plane. In the drawings of the movie moves, each side is represented by a portion of the *intrinsic sphere* —the sphere that is being mapped into space. The rest of the sphere dwells out of the bounds of the illustration. The folds, double points and triple points, however, can be drawn on a patch or several patches of the sphere. In the case of double points, there are two patches involved, and for triple points, there are three. As the fold and double point sets change, they sweep out surfaces in the time-elapsed sphere. The triple points sweep out arcs. After each illustration of a movie move, a corresponding illustration in patches of the intrinsic sphere is drawn. On the left side of the figures, the before and after views in patches of the ambient sphere are illustrated. On the right side, a set that is one dimension larger interpolates between the patches. For example, in the fold set, interpolating surfaces occur within a set of 2-spacial + 1-temporal-dimensions. Of course, $2 + 1 = 3$, and time becomes the left-right dimension of the figures.

In general, the *evolution* of an object refers to how the object changes within the deformations allowed. As the object changes, it is traced in a *time-elapsed* form. A time-elapsed sphere is the 3-dimensional set that consists of a thickened sphere — the material that forms a tennis ball or the peel of an orange. I could ascribe "before" to the inner white peel of the orange and the "after" to the orange side with time the interpolating quantity. But that metaphor interferes with the blue inside and red outside of the infinitesimally thin sphere. Instead, only patches of the sphere are considered at any one time, and for the duration of the time, the time-elasped patch is the 3-dimensional box with "before" to the left and "after" to the right.

The Fold Set

Go to the mirror, close your mouth, open your teeth, and poke your tongue into your left cheek. A fold appears that has an up-left cusp and a down-left cusp as its end points. The birth (or death) of a pair of cusps that are connected by a pair of fold lines is called the *lips* change because when the folds are drawn sideways, their introduction looked like a pair of lips. I think that the terminology is due to French mathematician, René Thom. To change a round red sphere into a round blue sphere, an arc of blue folds has to be introduced. One way of doing so is to introduce folds via the introduction of lips.

The affect to a movie and to the resulting surface is illustrated in Fig. 7.1.

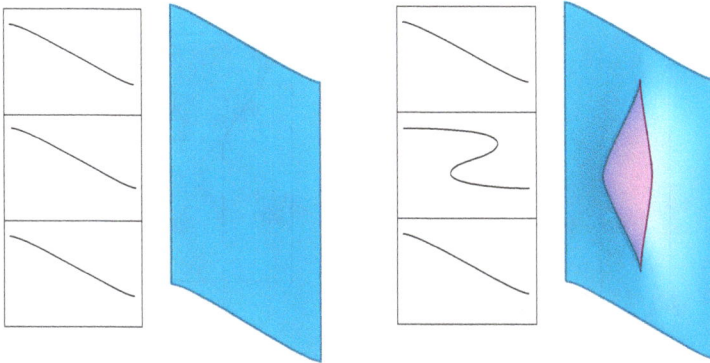

Fig. 7.1   The introduction or cancellation of a pair of cusps via the lips move

As in the discussion of dimensions (Chapter 4), the evolution of a cusp forms an arc in the time-elapsed picture; the evolution of a fold forms a surface which might have seams along its time-elapsed cuspal boundary. Therefore:

The evolution of the lips change forms a birth (or death) surface in the 3-dimension space that is the time-elapsed sphere. The surface of folds in this evolution has a seam formed by the cusp set. This seam of cusps is structurally similar to the set of folds of the original sphere, but you should envision it as the seam of a pita bread sandwich rather than the profile of an orange. In the sandwich, there is a separation between the two bread surfaces, they are joined at the seam, and the seam represents a sharp transition between the two surfaces. Similarly, in Fig. 7.2, the seam of the folds during the motion is illustrated by a blue line that separates the blue fold surface, which faces the viewer, from the red surface, veiled by the blue. This line is the evolution as the cusps are created from left to right.

———

Consider a down-pointing cusp and an up-pointing cusp; both of the same color and the same handedness (so the visible folds of both are on the left or both on the right) with the down-pointing cusp directly above the up-pointing cusp. These can be eliminated via the *beak-to-beak* cusp cancellation. Similarly, such a pair can be introduced when a visible fold

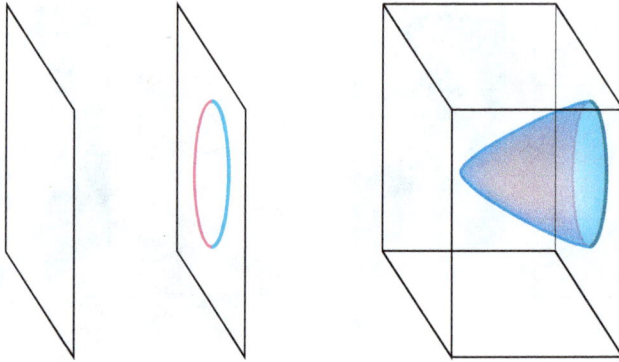

Fig. 7.2　The time-elaspsed evolution of lips within a patch of the ambient sphere

and a veiled fold of the opposite color and opposite handedness are nearby. Figure 7.3 illustrates the creation of the beaks from left to right.

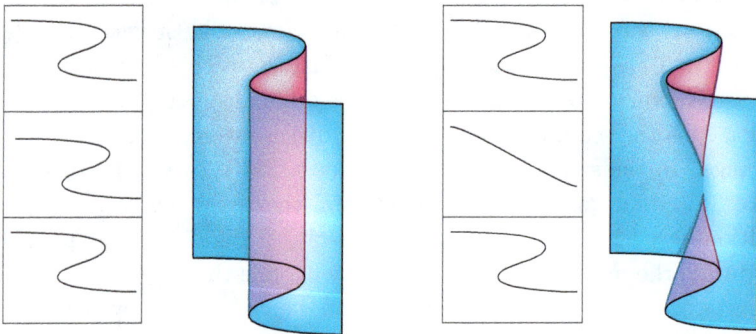

Fig. 7.3　The introduction or cancellation of a pair of cusps via the beak-to-beak move

The evolution of the beak-to-beak change in the fold set forms a saddle surface in the 3-dimensional space that is the time-elapsed sphere. The surface of folds during this evolution has a seam along the saddle. The seam is analogous to the seam of a pair of pants in a neighborhood of the crotch of the pants.

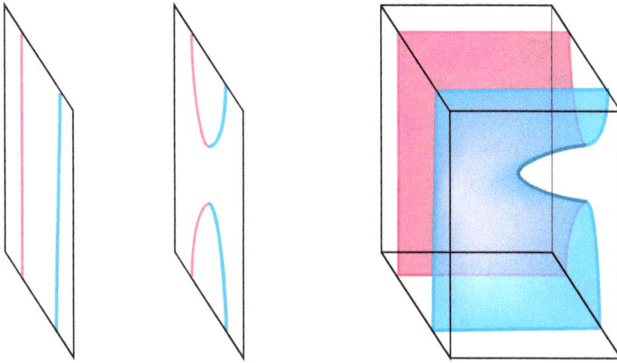

Fig. 7.4   The time-elapsed evolution of beaks within a patch of the ambient sphere

There is another way in which a pair of fold points can be introduced or destroyed. A simple fold arc can be interrupted via the introduction of a pair of oppositely colored cusps with differing handedness. The *swallow-tail* singularity occurs behind your knee when you bend or straighten your leg. Figure 7.5 illustrates the swallow-tail move.

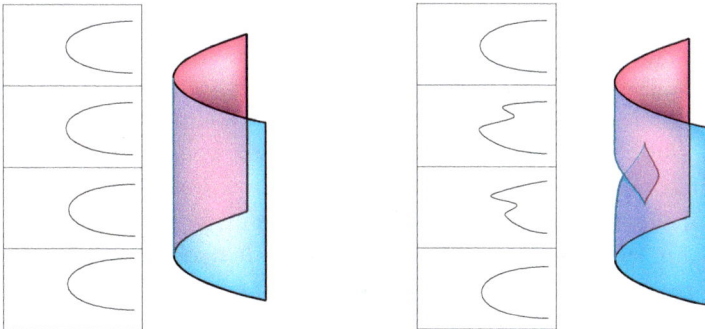

Fig. 7.5   The introduction or cancellation of a pair of cusps via the swallow-tail move

The evolution of the swallow-tail creates a cusp in the fold set of the time-elapsed sphere. On the left of the swallow-tail move, the fold set is a vertical line; say that it points down. On the right, the fold set (which originally pointed down and was homogeneously blue) now has a short red segment that points up followed by a second longer blue segment that points down. At the moment of the swallow-tail, the pair of opposing cusps are created, and this creation point is a cusp for the time-elapsed fold-set. Figure 7.6 illustrates.

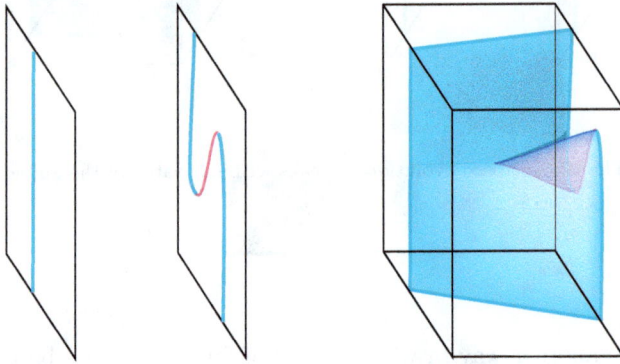

Fig. 7.6   The time-elaspsed evolution of a swallow-tail within a patch of the sphere

A fold line can bend. When it does, a canceling pairs of critical points are introduced in the movie. Specifically, a saddle point is introduced at the same time a birth or death is introduced. Figure 7.7 indicates the situation. Think of a mountain at the edge of a valley gradually eroding. In the cross-sectional movies, the canceling critical points are understood (from bottom to top) to be a saddle followed by the death of a simple closed curve. This is the only situation in which the fold set changes but there is no effect on the cusps. The move is called *critical point cancellation* or creation.

The evolution of the critical point cancellation between an optimum and a saddle point creates a cusp in the fold set of the time-elapsed sphere.

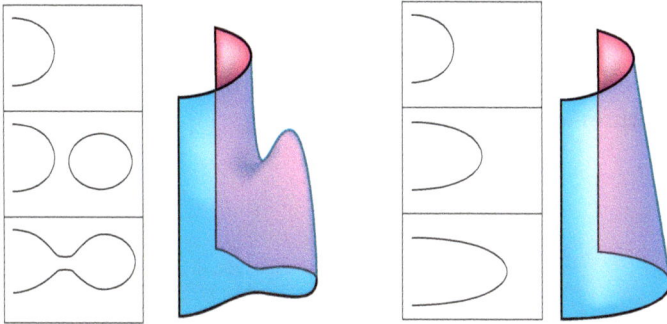

Fig. 7.7 Critical point cancellation or creation

Unlike the swallow-tail, the fold set is mono-chromatic. This situation is depicted in Fig. 7.8 where the optimal point is a maximal point, and this cancels with a saddle point from left to right.

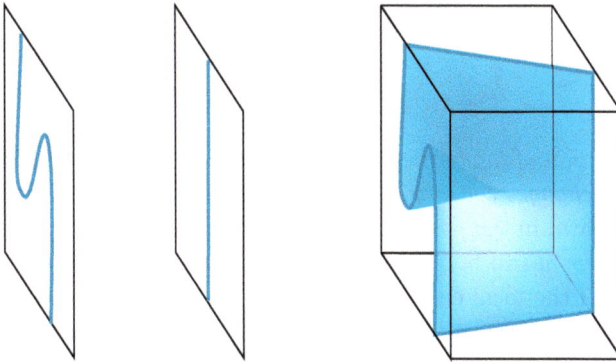

Fig. 7.8 The time-elaspsed evolution of the cancellation/creation of a saddle and optimal pair of critical points

In the last situation that involves only cusps and folds, the direction of a cusp changes in the presence of a saddle. The move is called a *horizontal cusp*. Figure 7.9 indicates a left-down-blue cusp changing to a right-up-blue cusp. In this move the color of the cusp remains constant but its up/down and left/right natures are interchanged.

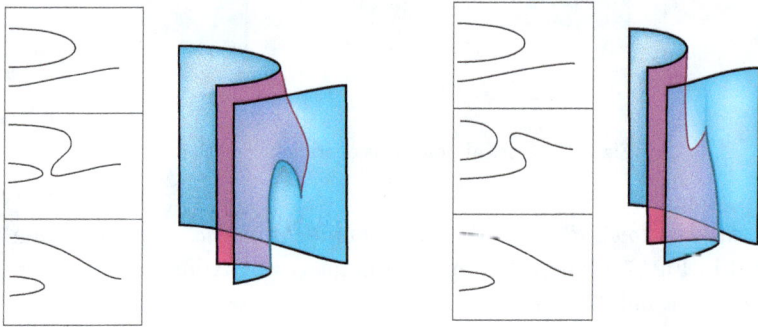

Fig. 7.9   A horizontal blue cusp

The evolution of the horizontal cusp involves a zig-zag in which the central segment changes color in the time-elapsed view. Figure 7.10 illustrates. An arch is drawn in the patch of the sphere because the critical behavior of the immersions has to be parroted on the intrinsic sphere in order to piece together all of the local changes. The need for the arch becomes apparent when the eversion proceeds. On the right of Fig. 7.10, the folds evolve to form a surface.

Observe that the creation/annililation of cusp via lips or the beak-to-beak move causes either an optimum or a saddle point to appear in the time-elapsed sphere. Similarly, the swallow-tail and the critical cancellation involves a type of cusp being created. The structures of optima, saddles and cusps repeat themselves in an increasingly baroque fashion throughout the processes. The duplications of form manifest analogies in their algebraic descriptions. Roughly, as the algebra becomes more complicated, within its deep inner structures, the familiar appear.

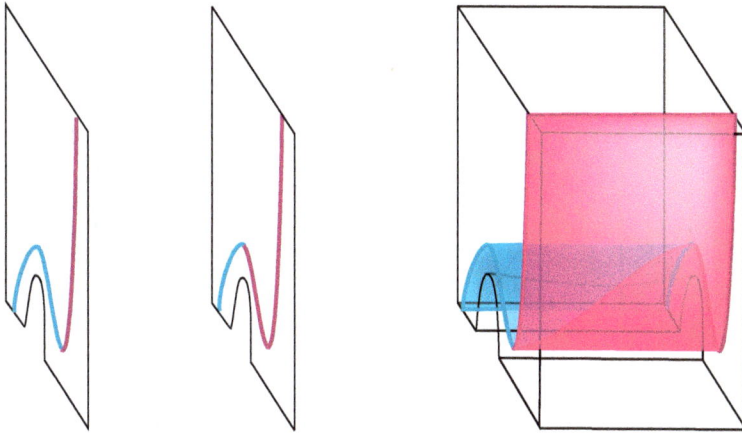

Fig. 7.10   The time-elaspsed evolution of the horizontal cusp

Double Points and Triple Points

A simple closed loop of double points can be created when two parallel sheets of the sphere touch and pass through each other. In the movie, a type II birth is immediately followed by a type II death. In the illustration a blue sheet pushes through a red sheet, forming a circle of double points. Figure 7.11 illustrates a *type II bubble move* or simply a *bubble move*.

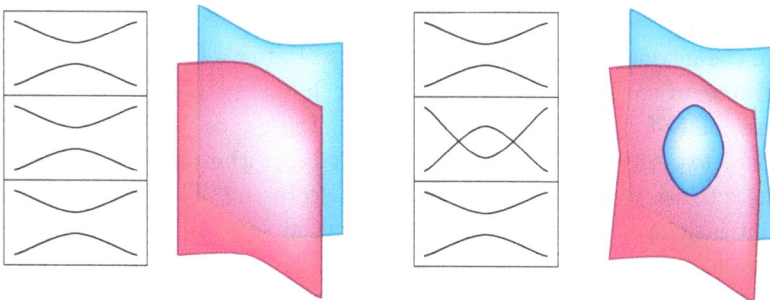

Fig. 7.11   A bubble move on the double point set

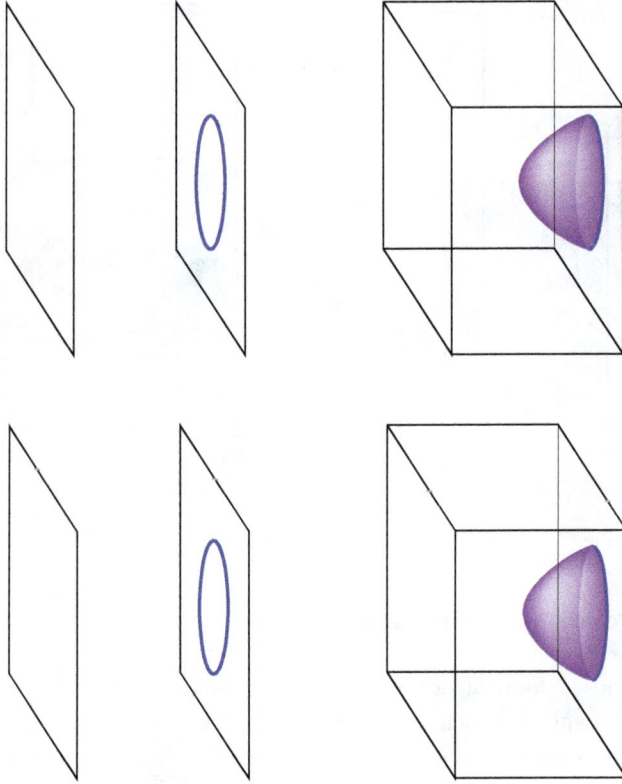

Fig. 7.12   The evolution of a loop of double points as it lifts to the intrinsic sphere

In the intrinsic sphere, two sheets intersect after a loop of double points is created in space. That is, each arc of double points in the immersed sphere comes from two arcs on the sphere. These arcs cover the arc that is in space two-to-one. The double point set lies within 3-space, and its lift to the intrinsic sphere is called the *double decker set*. The double points in the intrinsic sphere cover the double points in space like a double decker sandwich.

The evolution of the double point set in the bubble move forms a pair of bowls in the double decker set. These map in the obvious way to a single bowl in space. Figure 7.12 illustrates the birth of these bowls. The bubble move is, of course, reversible: A simple loop of double points that bounds a pair of disks on the sphere can be removed.

———

Consider a movie consisting of a type II death followed by a type II birth: A pair of crossing points between two arcs is replaced by two parallel arcs, and then crossings are reintroduced. Such a movie can be replaced by the pair of crossing points remaining static. Figure 7.13 illustrates the *type II saddle move*. On the immersed surface, a pair of parallel double point arcs that bound two strips are replaced by a pair of double point arcs in which the top has a minimal point and the bottom has a maximal point.

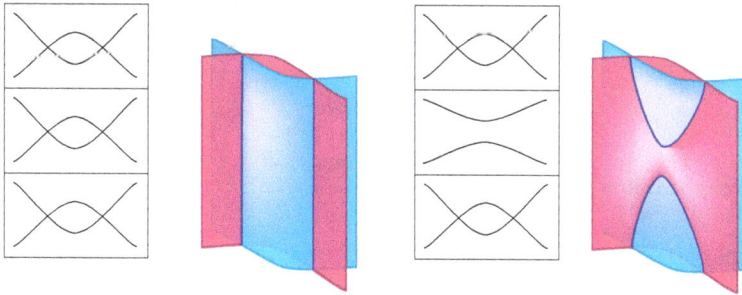

Fig. 7.13   A saddle move on the double point set

The evolution of the double decker set in a type II saddle move consists of a pair of saddles in the intrinsic sphere. Figure 7.14 illustrates. These two saddles cover a single saddle of double points in (3+1)-space-time.

———

Consider a movie in which the first still contains a pair of arcs that cross once. The second still is obtained from the first by a type II birth *to the right* of the original crossing, and the last still is obtained from the second by the removal of the two crossings *on the left* via a type II move. This movie can be replaced by a sequence of three stills in which the crossing

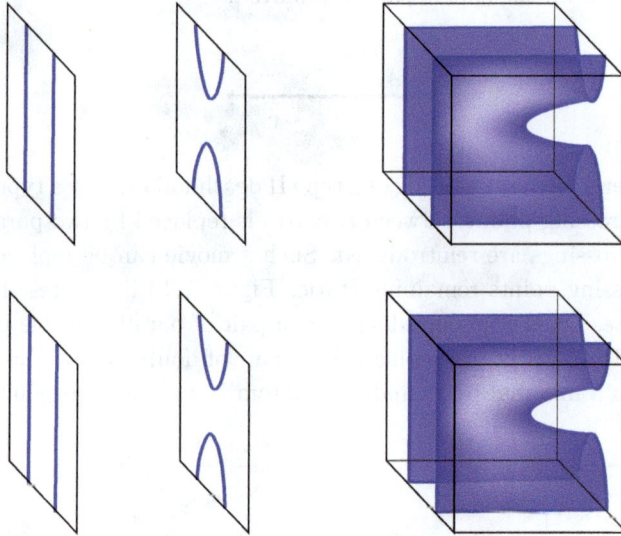

Fig. 7.14   The evolution of a pair of arcs in double decker set in the intrinsic sphere during a type II saddle move

remains virtually still and nothing further happens. Figure 7.15 illustrates the *type II zig-zag move.*

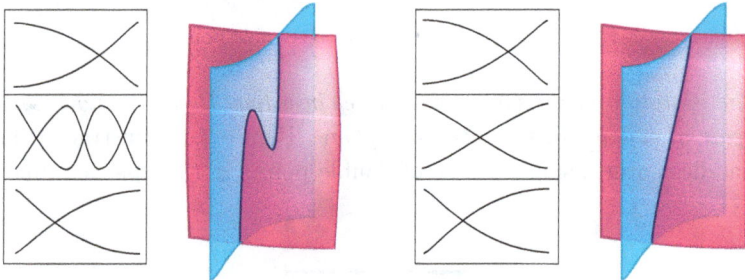

Fig. 7.15   A type II zig-zag move

The evolution of the double decker set in a type II zig-zag move consists of a pair of cusps on the double decker surface. Figure 7.16 illustrates. As

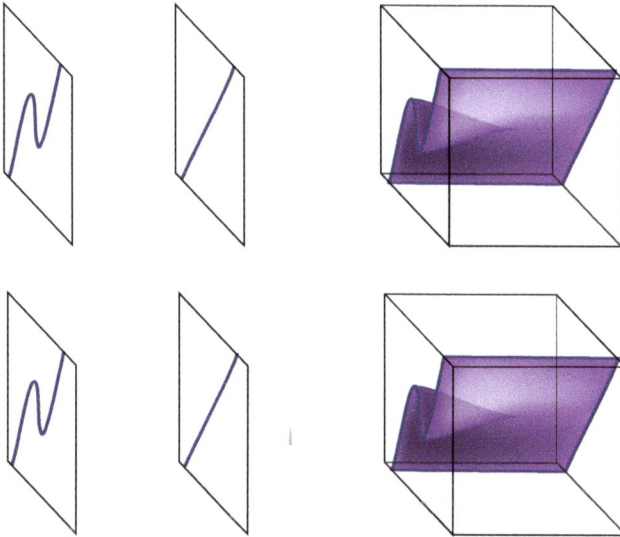

Fig. 7.16   The evolution of the double decker set in the intrinsic sphere during a zig-zag move

in the cases of the type II bubble and type II saddle moves, these cusps map to a single cusp in the double point set in $(3+1)$-space-time.

———

The *type III-type III move* involves replacing a pair of type III moves by an inertial situation. Figure 7.17 illustrates.

The evolution of the triple decker set in a type III-type III move involves three copies of a type II birth. Figure 7.18 illustrates. The triple points of the sphere are double points of the double decker set. At a fixed moment, an isolated triple point lifts to the intrinsic sphere to three points. Each of the three double point arcs that cross at the triple point, lifts to a pair of arcs in the double decker set. So the *triple decker set* for an isolated triple point consists of three copies of a crossing between two arcs. In the type III-type III move, a pair of triple points are created from left to right. The six double decker arcs involved occur in three pairs, and proceeding left to

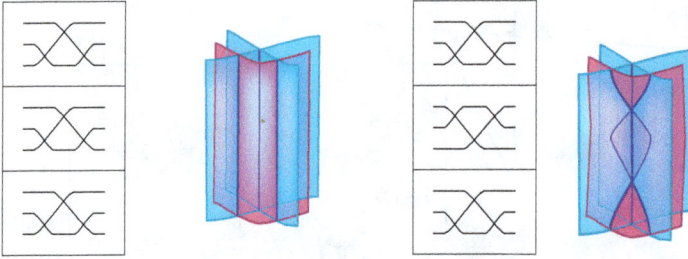

Fig. 7.17   A type III-type III move

right these change from being three pairs of parallel arcs to three pairs of arcs with two crossing points each.

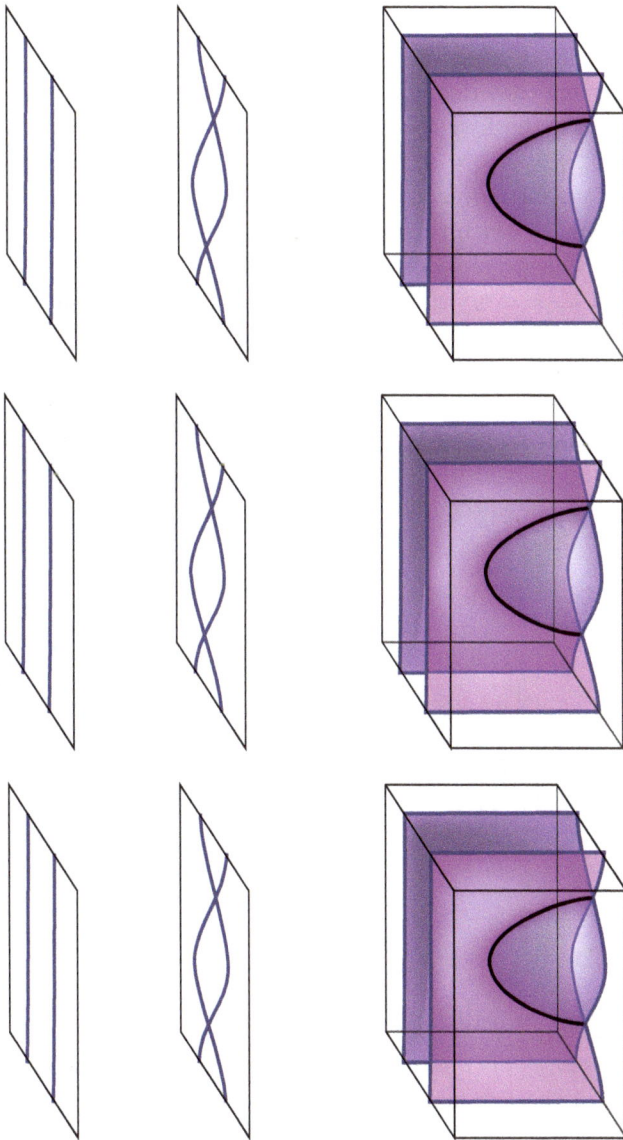

Fig. 7.18 The evolution of the triple decker and double decker sets in a type III-type III move

The *quadruple point move* or tetrahedral move is the most important move from the point of view of this sphere eversion. In this situation, four sheets of the sphere form a tetrahedron whose four vertices are triple points. We consider one of the sheets to pass through the other sheets thereby inverting the orientation of the tetrahedron. There is a coordinate system in which the four sheets of the sphere can be identified with the planes $x = 0$, $y = 0$, $z = 0$, and $x + y + z = 1$ before the move, and for which the last plane moves to $x + y + z = -1$ at the end of the move. At the singular instant at which this plane passes through $x + y + z = 0$, the four triple points coalesce to be a quadruple point. According to a Theorem of Banchoff and Max, every sphere eversion has at least one quadruple point. Many eversions (for example, Outside-In) have a singular point at which more than four sheets converge. The immersion that I illustrate has exactly one quadruple point, and therefore, we can separate the eversion into a red side or a blue side. Each immersed sphere that we see is either red or blue depending on which side of the quadruple point the immersion lies. Figure 7.19 illustrates the quadruple point move.

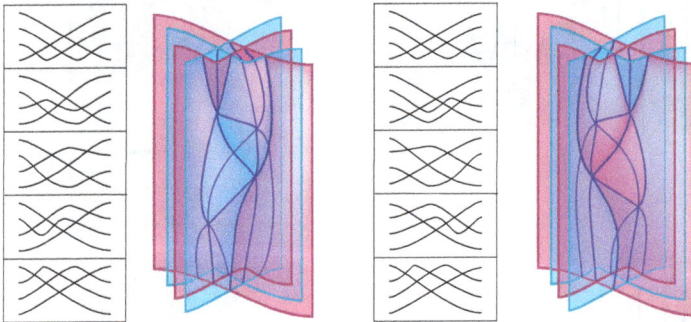

Fig. 7.19   A quadruple point

The evolution of the decker set under the quadruple point move involves four type III moves among the twelve triple decker arcs. Label the arcs on the top still of the left hand movie 1 through 4 from left to right. Each patch of the sphere before the move can be thought of as a rectangle labeled 1 through 4. In the first sheet, there are three arcs of double points that can be indicated by 2, 3, and 4. The triple points on sheet 1 appear as double points among these arcs. From top to bottom they are (23), (24), and (34) so that they correspond to the triple points (123), (124), and (134) on the first sheet. The notation (123), for example, indicates the intersections among the sheets labeled 1, 2, and 3. On the left side of the quadruple point movie, the triple points from top to bottom are (123), (124), (134), and (234). The bottom most triple point does not occur on sheet 1 and therefore on the left there are three straight segments of double points at the bottom of that sheet. On the right from top to bottom, the triple points are (234), (134), (124), and (123). The same triple points appear in the opposite order. So on the right on the first sheet from top to bottom, the intersections are (34), (24), and (23) — the same sequence as on the left but in the reverse order. The difference between right and left is achieved by a type III move.

The analysis of the previous paragraph applies to each of the four patches of the sphere involved with the labeling adjusted, of course. The quadruple point is the simultaneous occurrence of four triple points of the double decker set.

Double Points and Folds

The double point set can interact with the fold set and the cusp set in several ways. This section describes the movie moves in which the relationship between the double point set and the fold set changes.

The ψ, ψ-*move* replaces a double point arc in the neighborhood of fold with an arc that bounces back and forth over the fold. Figure 7.21 illustrates.

The evolution of the decker set under a ψ,ψ move is a type II move between a fold and one of the double decker arcs with the other acquiring a slight bend. One of the double point arcs before the move lies parallel to the fold line on that patch of the sphere. The other arc lies on a fundamentally straight plane. After the move, the straight plane has bent a little, and the double point arc has a segment on the other side of the fold. Figure 7.22 illustrates.

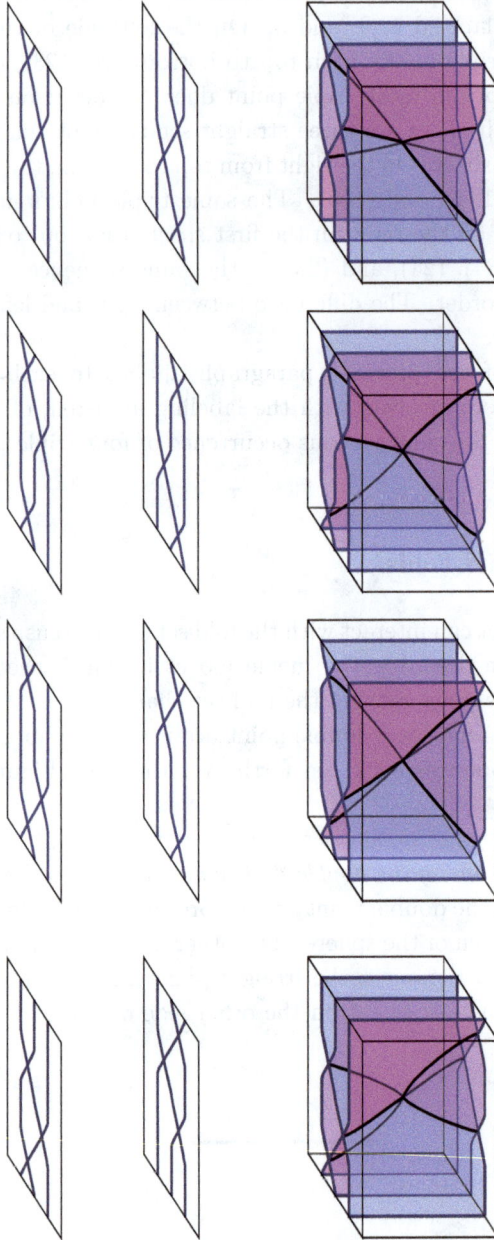

Fig. 7.20   The evolution of the decker set under a quadruple point move

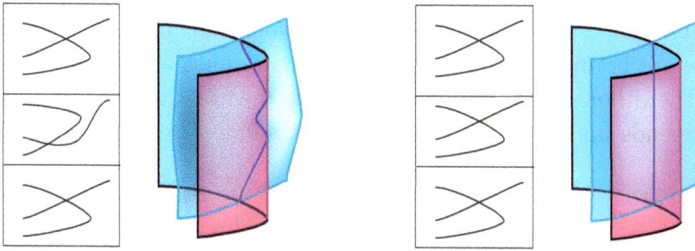

Fig. 7.21   A ψ, ψ-move

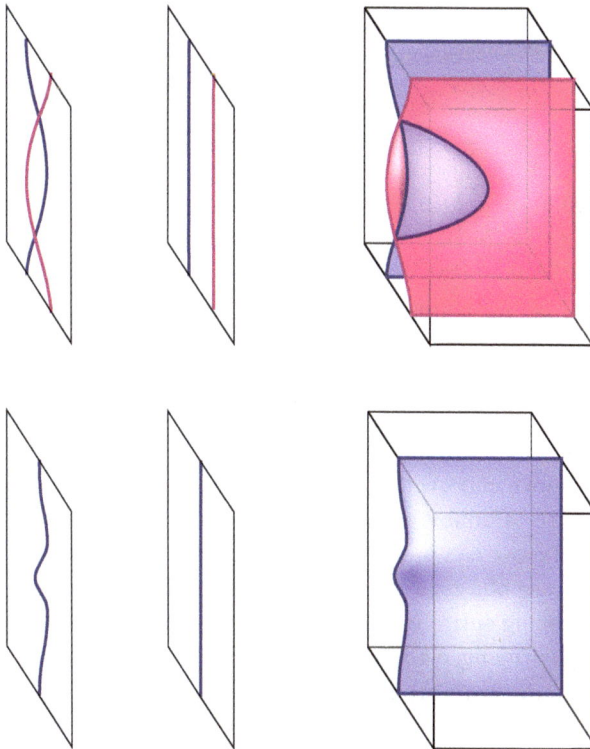

Fig. 7.22   The evolution of the decker set under a ψ, ψ move

A *double point bouncing through a cusp* causes the proximity of a cusp to the observer to change when one of the folds that ends at the cusp has a double arc bouncing over it. The double arc must also cross in front or behind the other fold that defines the cusp. After the move, the double point arc ends up bouncing over the other fold. Figure 7.23 illustrates.

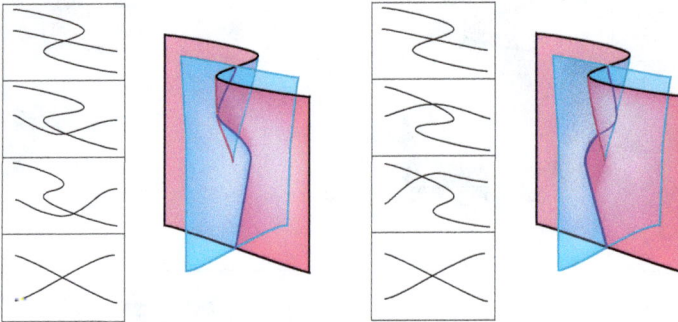

Fig. 7.23   A double point bouncing through a cusp

The evolution of a double point bouncing through cusp in the decker set consists of a double point arc bouncing over a cusp line of the fold surface in one sheet, and the other sheet of the double point surface gaining a slight bend. Figure 7.24 illustrates.

A type II move can occur in a region near a maximum, minimum, or saddle. If an arc of double curves bounces over one of the resulting folds, then the type II move can pass over the optimum on the fold set. Figure 7.25 illustrates the case of a *type II move passing over a maximum* with the case of a minimum being entirely analogous. Figure 7.26 illustrates the case of a *type II move passing over a saddle*. This picture can also be turned upside-down. These moves are grouped together within this paragraph because from the points of views of the drawings, they appear to be quite similar.

Both an optimum and a saddle induce an optimum on the fold set. The birth of a circle induces a minimum, $\smile$, the death of a circle induces a maximum, $\frown$, and a saddle induces either type of optimum on the fold set. There are two easy ways to quantify the differences between an optimum

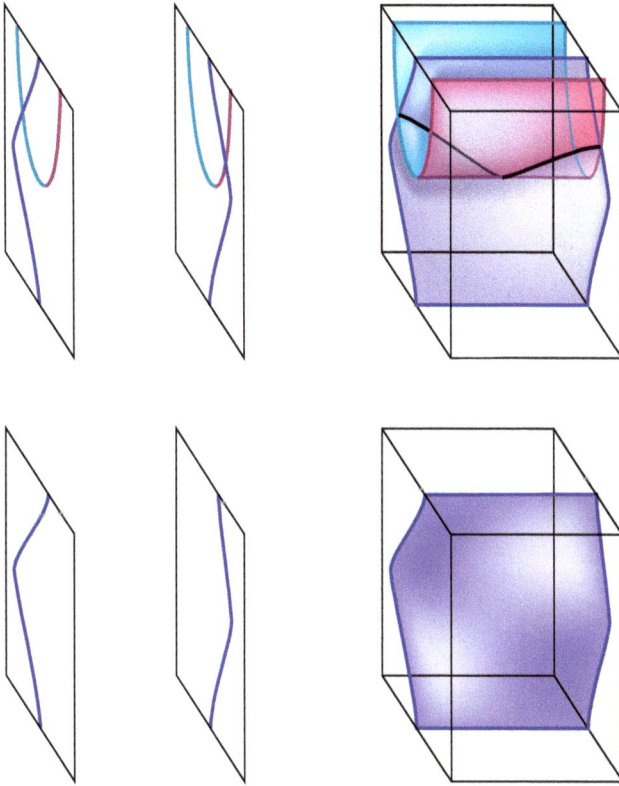

Fig. 7.24   The evolution of the decker set when a double point passes through a cusp

and a saddle. At a saddle point the ⊃ (or maximum) is to the left of the ⊂ (or minimum) while at a birth or death, the left-pointing ⊂ is to the left of the right-pointing ⊃. Thus, the surfaces, that converge at the fold for the saddle, project to the outside of the ⌣ or ⌢ while they project to the inside at births and deaths.

The evolution of the decker set when a type II move passes over an optimum or a saddle consists of a ψ move with a sheet of the fold surface bouncing over a sheet of the double point surface while the other double point surface forms a parabolic arch or trough. Figure 7.27 illustrates one possible case.

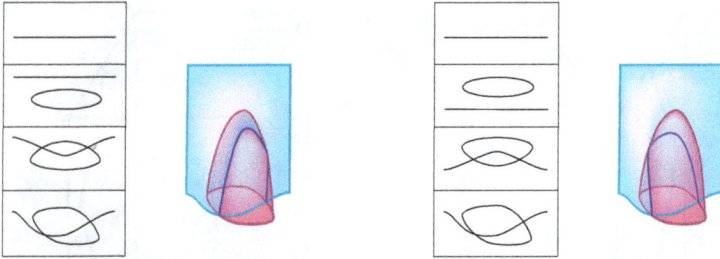

Fig. 7.25   A type II over an optimum

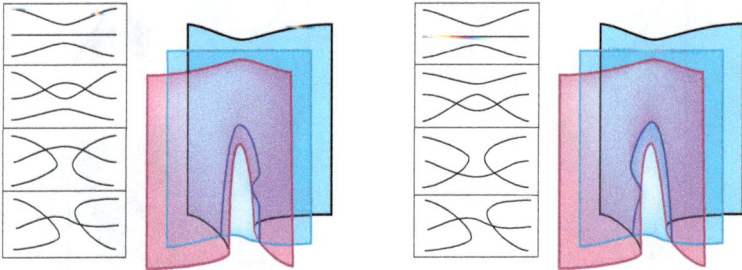

Fig. 7.26   A type II moves over a saddle

The *horizontal type II move* is among the more difficult to illustrate even though from some points of view, the move is quite natural. In it, a type II move is to occur when two optima within the stills of the movie are present. In the given illustration, the optima are both minima ($\subset$), but they could both be maxima ($\supset$), or one of each type. In fact, a clever application of the other moves makes any two of these three possibilities a consequence of the remaining one, but that subtlety does not concern us here. By convention, a type II birth occurs by taking a pair of parallel horizontal arcs and introducing a pair of crossings in these arcs. When

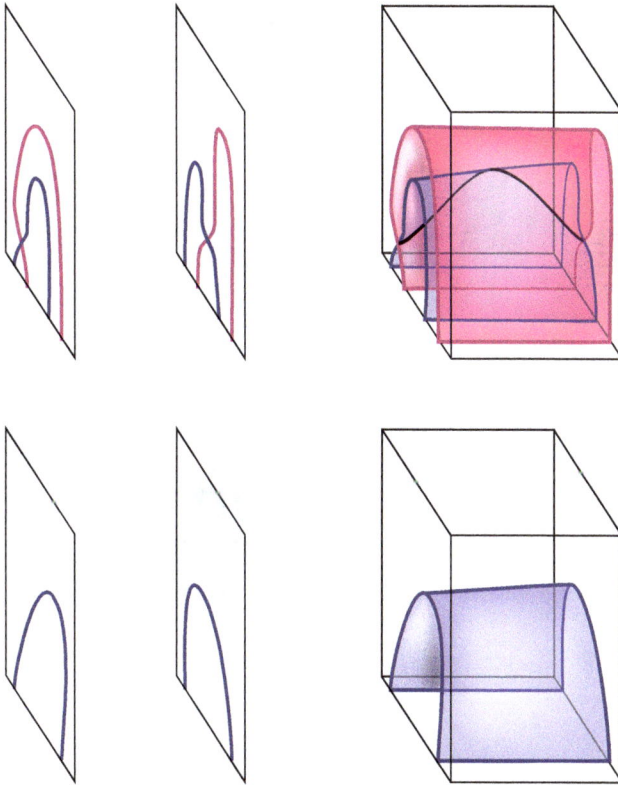

Fig. 7.27 The Evolution of a type II move over a saddle or an optmimum

the arcs are not horizontal but have optima on them, the optima have to bounce out of the way for the move to occur. There is a choice of which optimum moves first, and the horizontal type II move reflects the order of these choices. Figure 7.28 illustrates the move.

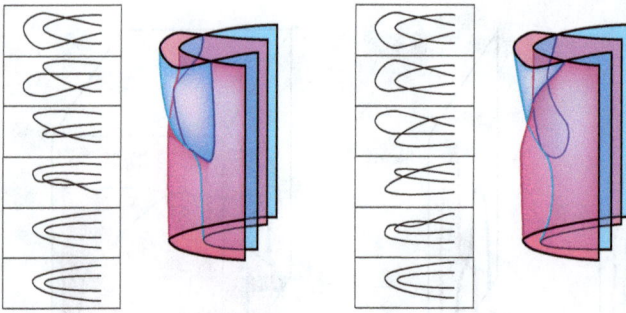

Fig. 7.28   A horizontal type II move

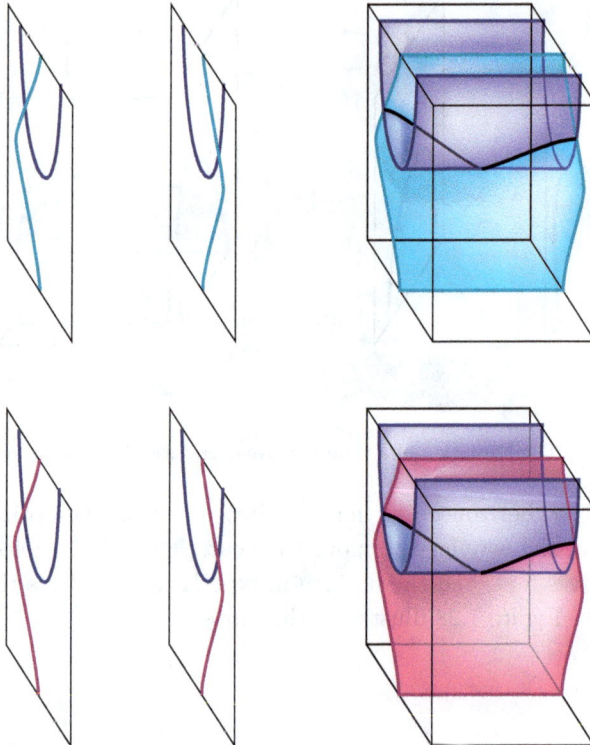

Fig. 7.29   The evolution of the decker set under the horizontal type II move

The evolution of the decker set under the horizontal type II move consists of a pair of ψ bounces with the fold surface bouncing over the optimum of the double point set. Figure 7.29 illustrates.

Triple Points, Double Points, and Folds

Observe that a type II birth or death is a critical point (minimum or maximum) for the height function of the sphere when it is restricted to the double point set. If there is a triple point nearby, there is a corresponding ψ-move between the triple point and the type II move. In this way, the triple point can bounce to the other side of the optimal point on the double point set. Figure 7.30 illustrates one of the variations of the move called the *triple point bounce*. You can easily determine the other variations of this move.

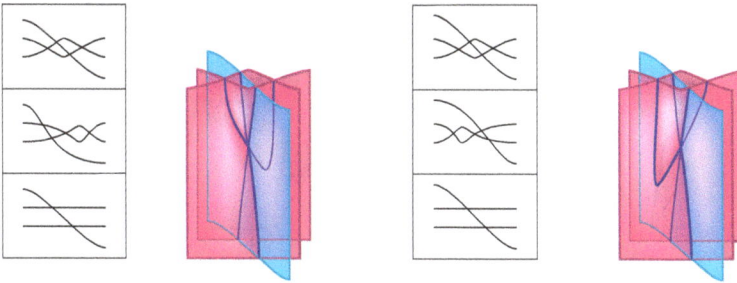

Fig. 7.30   A triple point bounce

The evolution of the triple decker set in a triple point bounce consist of three copies of the corresponding double point bounce. Figure 7.31 illustrates. Again the triple decker set is considered as the points of intersection in the double decker set as these sit in the intrinsic sphere.

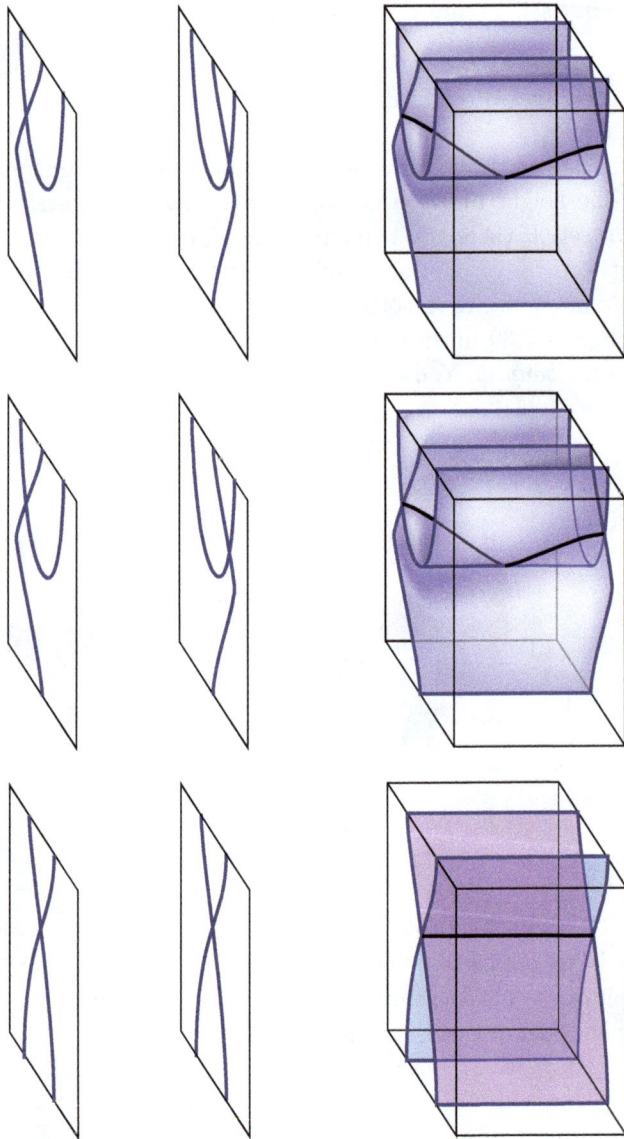

Fig. 7.31 The evolution of the triple decker and double decker sets in a triple point bounce

In a situation similar to the horizontal type II move, a type III move may be obstructed by an optimum point. The type III move is supposed to occur in the presence of a triangle, but one side of that triangle may be bent. In a *type III passes over a fold* move, the fold bounces either in front or behind the triangle over which the type III move is to occur.

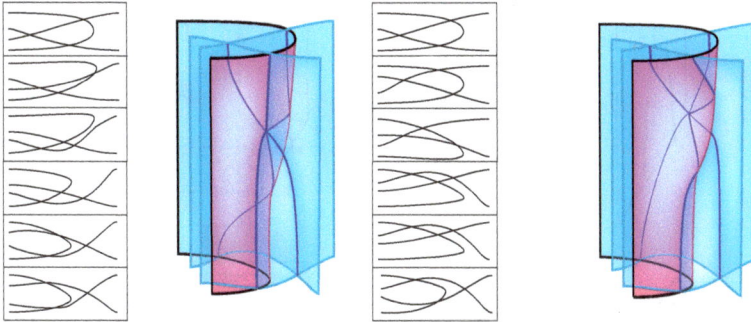

Fig. 7.32   A triple point passes over a fold

The evolution of the decker set under a triple-point-passing-over-a-fold consists of a triple point between the fold line and two of the double decker sheets while the other four double decker sheets continue to cross as pairs of generic planes. Figure 7.33 illustrates.

There are a few more situations that are not explicitly articulated, but are part of moves to surfaces. Critical levels in the charts may exchange positions provided they lie far enough away from each other. Also, the exchange of vertical positions within the charts happen without much fanfare.

Within the movie context, it is quite easy to confuse folds that both point left ($\subset$) or both point right ($\subset$), and mistakes occur when such folds are confused. Labels as developed in the Gauss-Morse section of Chapter 10 make things more clear.

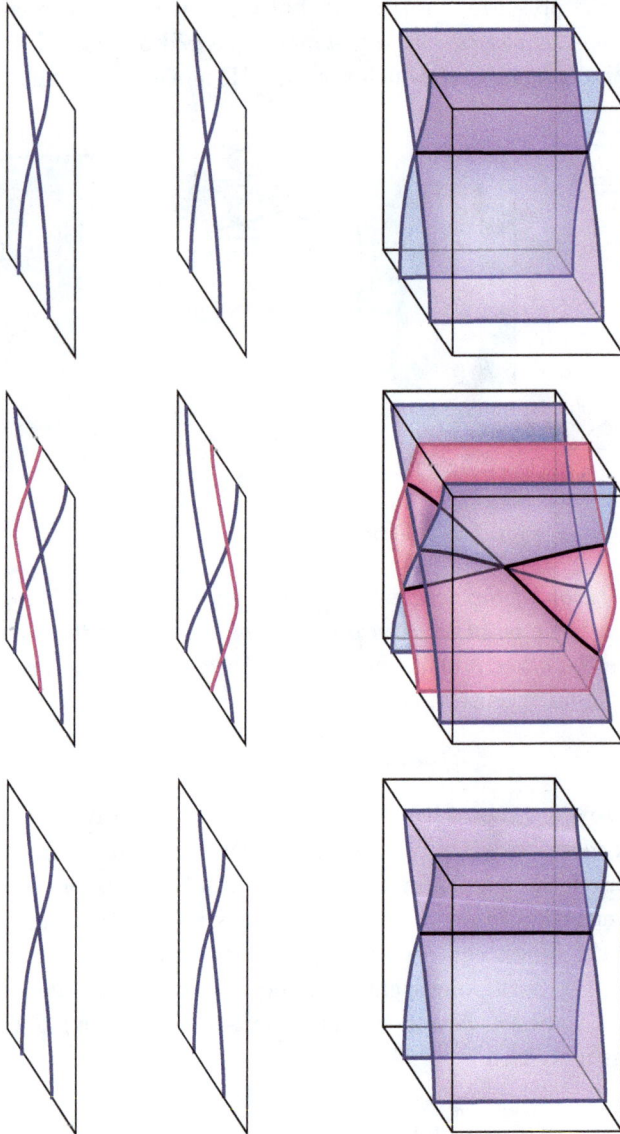

Fig. 7.33   The evolution of the decker set as a triple point passes over a fold

Conclusion

As you can see, the description of the movies moves and their effects on the decker sets requires patient, vigilant, and visual intuition. The evolutionary results upon the decker sets mimic the pieces that constitute the immersed sphere. As the sphere evolves, each step is described by a movie move taken from the set given here. In rare situations, a sequence of moves occurs at a single step. To recapituate the ideas of this chapter, the next chapter is a tabulation of the dimensions of the resulting sets (double point set, triple point set, double decker set, *etc.*).

Chapter 8

Taxonomic Summary

An immersed sphere is described by a movie. Any movie is decomposed as a sequence involving some of (or all of) the following scenes: birth, death, saddle, cusp, type II, ψ, critical exchange, or a type III move. A height function (left-to-right) on the stills of the movie allows the critical points ($\supset$, X, and $\subset$) of one still to be traced to the next, and subsequently throughout the scene. The scenes, then, give rise to optima on folds, cusps, births and deaths of double points, double points passing over folds, and triple points. These are, in turn, the critical points with respect to the height function (down-to-up) of the corresponding illustration of the immersion. The changes in critical behavior, as the immersion changes, are tracked via the movie moves which quantify singular immersions. The singularities occur at the blink of an eye and, therefore, are depicted by a before/after point of view. The singularities are caused by coincidences among the critical points in the illustrations. It is a good moment to step back from the previous two chapters and to summarize the critical points and the singularities in a tabular format. Doing so rounds out Chapters 4, 6 and 7 by providing an encapsulated glossary.

Scenes that constitute a movie	
Scene	Result
Birth/Death	Smooth creation/annihilation of a monochomatic pair of folds with min. on the left and Max. on the right
Saddle	Smooth creation/annihilation of a monochomatic pair of folds with Max. on the left and min. on the right
Cusp	Sharp creation/annihilation of a pair of folds with min. on the left and different colored Max. on the right
Type II Birth/Death	Creation/annihilation of a pair of double points
Type III	Convergence and subsequent divergence of three double point arcs; in the before and after stills, the three double points bound a triangle
ψ	Double point arc and fold line exchange depths
Critical exchange	A pair of distant critical points exchange left and right positions

Within the encoding of the fold lines, folds that come from $\subset$ are labeled with lower case letters r for red, and b for blue. Similarly, folds of the form $\supset$ are labeled R for red and B for blue. As a further affectation, the word *minimum*, its plural (minima), and its abbreviation (min.) begin with lower case letters while *Maximum*, Maxima, and Max. begin with upper case letters. The notation r, b, R, B is explained in more detail in Chapter 10 where Gauss-Morse codes are explained.

Movie-moves (part 1)		
Movie-move	Result	Time Elapsed
Lips	Creation/annihilation of a pair of cusps connected by oppositely colored folds	An optimum on the fold set
Beak-to-beak	Creation/annihilation of a pair of cusps by breaking or joining a pair of fold arcs	A saddle on the fold set
Swallow-tail	Creation/annihilation of a pair of cusps by interrupting a fold of one color with a segment of the other color	A cusp on the fold set
Horizontal cusp	The (up/down) direction of a cusp changes as a saddle turns upside-down	The bench of folds changes color

Movie-moves (part 2)		
Movie-move	Result	Time Elapsed
Type II bubble	Annihilation/creation of simple loop of double points	Cap/bowl in the double point set
Type II saddle	Creation/annihilation of a min. over a Max. of the double point set via a saddle	Saddle in the double point set
Type II zig-zag	cancellation/creation of a Max./min. pair of type IIs	Cusp in the double point set
ψ, ψ	cancellation/creation of a pair of double points passing over a fold	Death/birth on the set of bounce points

Movie-moves (part 3)		
Movie-move	Result	Time Elapsed
Type III-Type III	cancellation/creation of a pair of triple points	Arc with an optimum in triple point set
Quadruple point	Interchange of the position of 4 triple points that form a tetrahedron	Quadruple point at the singular moment
Type III bounce	Move an optimum on the double point set to the left/right of a triple point	ψ move between triple point and type II

Movie-moves (part 4)		
Movie-move	Result	Time Elapsed
Type II over optimum	ψ moves to other side of the optimum	ψ move between bounce point and optimum
Type II over saddle	ψ moves to other side of the saddle	ψ move between bounce point and saddle
Type II over cusp	cusp moves to other side of a veil	ψ move between bounce point and cusp
Type III over fold	Triple point moves to other side of the veil formed by folded sheet	apparent tangency between triple point arc and fold surface
Horizontal type II	Type II move in the presence of a pair of folds	Branch point within the double point surface

Next, I tabulate the sets of points that are of interest, and remind you of their dimensions at each time, and as the process evolves.

Dimensions of the sets		
Set	Fixed Time Dimension	Time elapsed Dimension
Cusp	0	1
Folds	1	2
Max./min. & saddles on fold set	0	1
Double points	1	2
Triple points	0	1
Bounce points	0	1
Max./min. & saddles on double point set	0	1
Quadruple point	–	0
Intrinsic Sphere	2	3
Ambient space	3	4

Of course, every number in the third column is one more than the number in the second column since the evolution of time causes dimensions to increase by one. There is only a quadruple point at a singular instant, so its dimension at any fixed time does not make sense. Finally, a 0-dimensional point set is a set whose points can be separated from each other. At any time, there may be as many as four triple points, but they are distinct. Meanwhile, the double points and folds form arcs at any fixed time, and therefore cannot be separated.

Chapter 9

How Not to Turn the Sphere Inside-out

The disallowed scene in all of our movies is a branch point (see Fig. 5.3). It is called a *type I move* — you no longer need be curious what comes before a type II move. It is disallowed since the image of a sphere that has branch points has at least one pair of these and at these points there is no well-defined tangent. The folds change colors at the branch points. As a warm-up to the real thing, I show you a method of turning a sphere inside-out which includes branch points. But the process is too easy — like ending a game of Risk by kicking over the board. However, the movie, picture, and decker sets here are quite a bit more simple than in the real eversion. So it serves as a good model to see how processes work. Here goes.

Starting from the red sphere (Fig. 9.1), a pair of branch points is introduced (Fig. 9.3) via a bubble move of branch points which is not discussed in detail here. The branch points move over the maximum and the minimum between Figs. 9.3 and 9.5. Then they cancel via a saddle move on branch points between Figs. 9.5 and 9.7. This sets up a type II bubble move that pushes the sphere to be blue between Figs. 9.7 and 9.9. Find the corresponding decker sets illustrated in Figs. 9.2, 9.4, 9.6, 9.8, and 9.10.

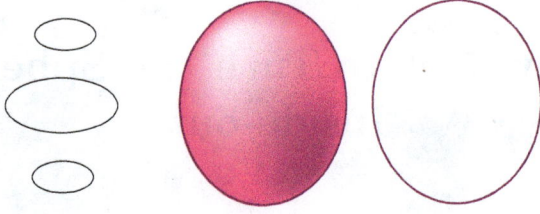

Fig. 9.1   A red sphere and its movie

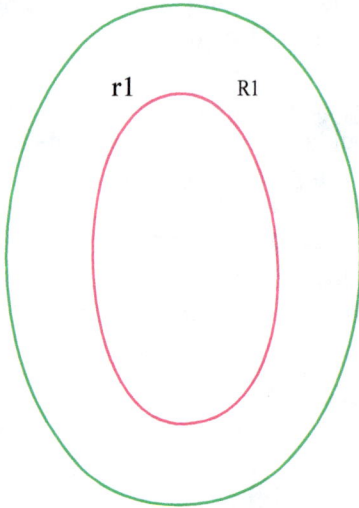

Fig. 9.2   The first decker set in the branch point movie

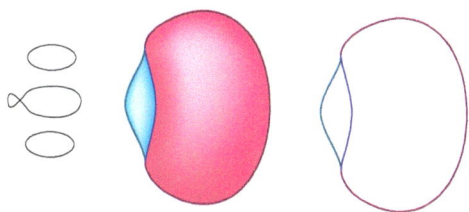

Fig. 9.3   A pair of branch points has been introduced

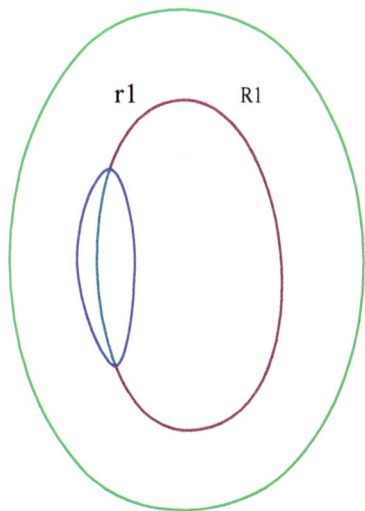

Fig. 9.4   The decker set after branch points are added

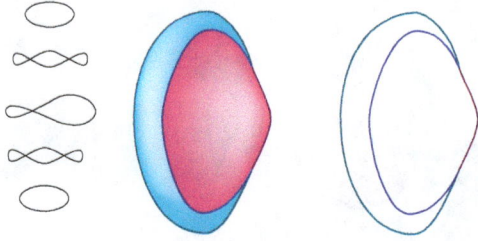

Fig. 9.5   The branch points have moved over the optima and to the right

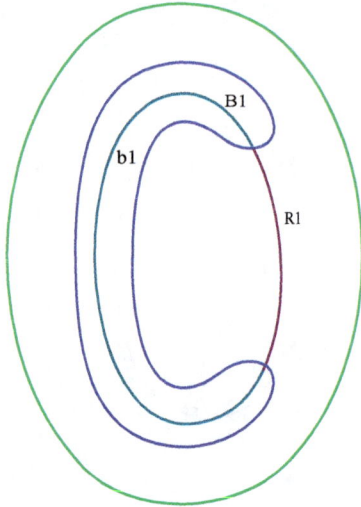

Fig. 9.6   The decker set with branch points on the right

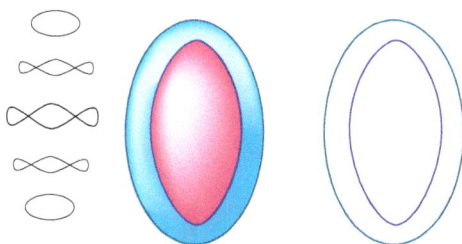

Fig. 9.7   The branch points have canceled

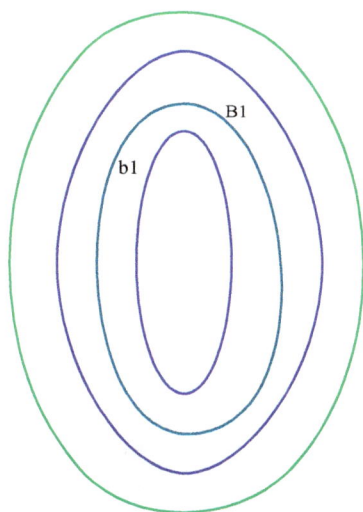

Fig. 9.8   The decker set after the branch point cancellation

Fig. 9.9   A type II bubble has turned the sphere blue

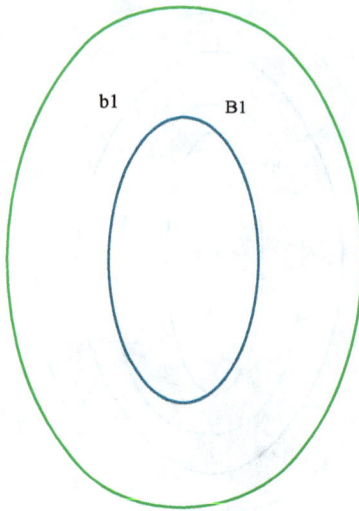

Fig. 9.10   The decker set of the blue sphere

Chapter 10

A Physical Metaphor

Throughout recorded history, mathematics has been used to explain the natural world. Mathematical models can be physical or they can be metaphysical. That is, mathematics also models hypothetical universes. The mathematician often engages in a mind game: If the world were of some specific form, the consequences would be what? In this chapter, I describe a metaphysical world in which particles interact along a line, and use that world to construct a universe of 1 spacial dimension and 3 temporal dimensions. At the end of the day, the sphere eversion is possible in this world. Meanwhile, the particle interactions, the interactions among interactions, and the interactions among these, form an algebraic system.

Consider a universe that consists of a vertical line. At any time in this world, there are a finite number of positively and negatively charged particles. We want balance in this world, so at any time, the number of pluses coincides with the number of minuses. There are three basic types of interactions that occur in time: a pair of oppositely charged particles are born next to each other, a pair of oppositely charged particles are annihilated by each other, and any pair of particles can exchange positions.

Time is schematized to proceed from left to right. The birth of a pair of particles in which the positive particle is on top is given by $\subset^+_-$. If the negative particle sits on top, then the symbol is $\subset^-_+$. The death of a pair of particles in which the positive particle is on top is given by $^+_- \supset$, and if the negative is on top, the symbol is $^-_+ \supset$. The interchange of particles is schematized as one of:

- positive particles interchanging position: $^+_+ X^+_+$
- negative particles changing position: $^-_- X^-_-$
- a negative above a positive changing to a positive above a negative: $^-_+ X^+_-$

- a positive above a negative changing to a negative above a positive: $^+_-X^-_+$

The time-like traces of the motions of such particles are directed line segments, or more conveniently, arrows. A negatively-charged particle, at a specific time, is the end point of left pointing arrow towards the past ($\leftarrow \cdot$) or away from the future ($\cdot \leftarrow$). A positively-charged particle is the end point of a right pointing arrow ($\cdot \rightarrow$ or $\rightarrow \cdot$). The possible universes occur in $(1+1)$-*space-time*, or a *space-time continuum*.

Such universes consist of oriented arcs and circles in the plane. Let us look at some examples. The most simple of all universes is the one in which no particles are born nor do any die. It consists of a rectangular strip in the plane devoid of motion and interaction. The next most simple universes are described by $\subset^+_-\supset$ and $\subset^-_+\supset$. The former is a clockwise oriented circle and the latter is counterclockwise oriented circle.

In the possible universes imaginable, I want to declare a conservation law: *The number of particle exchanges (X) is always even.* So it may happen that no particle exchanges exist or there may be several, but there are always an even number.

What is the set of all possible space-time universes that start from nothing and end at nothing? It is the set of immersed oriented closed curves (with possible many distinct circles being immersed) that have an even number of double points and that have a fixed ordering on its set of critical events. These events are represented by $\subset$, $\supset$, and X. Any still in a movie of our eversion represents one such universe.

The number of these worlds is, in fact, infinite. To better understand these universes, let us consider how they may be related. For example, it is possible that certain critical events can occur simultaneously. There are two perturbations of such universes in which one event proceeds another, or *vice versa*. In the $(1+1)$-space-times, if two events could occur simultaneously, then we might consider the two perturbations, one before the other or the other before the one, to be equivalent. The equivalence between these universes is represented by one of the critical exchanges which are represented as possible movie scenes. As another example, the exchange of position of a pair of charged particles followed by their exchange back, may be considered to be inconsequential. So an omniscient being observing two such worlds (one in which a pair of particles remained fixed in the firmament and the other in which the particles switch back and forth) would understand the connection between these worlds.

You, the omniscient observer, might observe some other local relationships among possible universes. These (and the ones above) are related by the scenes: critical point exchange, type II, type III, cusp, and ψ-bounce. In other words, these (somewhat natural) relationships between $(1 + 1)$-dimensional continua correspond to some of the scenes in a movie. The missing scenes (birth, death, and saddle) also correspond to relationships among universes which have a somewhat different nature.

Specifically, the space-time that consists of an oriented (clockwise or counter-clockwise) circle corresponds to the creation of a pair of particles followed by their annihilation sometime later. That universe has substance for some finite period of time, yet the empty universe is a void. Those two universes seem quite a bit different, but are related by the birth or death of the circle. George Bailey, the Jimmy Stewart character in "It's a Wonderful Life," realizes that the world with him in it is qualitatively different from the world without him in it. The two $(1 + 1)$-dimensional universes — one without the specific life of a pair of particles and the other in which these particles are born, interact, and die — are substantially different.

The saddle relationship in the context of signed particle interactions needs some explication. Consider a $(1 + 1)$-space-time that contains the interaction $\overset{+}{\underset{-}{\supset}} \subset^{-}_{+}$. The saddle relationship converts that to $\overset{\leftarrow}{\rightarrow}$. Such a saddle preserves all of the orientations of the resulting curves. Also, it has an interesting effect with respect to the connectivity of the curves: An oriented saddle relationship disconnects a single closed curve into two oriented curves, or it connects two curves into one.

Another context happens in which the relationships: birth, death and saddle are significantly different than the relationships: cusp, type II, type III, ψ-bounce, and critical exchange. The latter relationships satisfy involutive movie moves. For example, a movie that has a type II scene that creates a pair of double points followed by their immediate annihilation can be replaced by the movie which is static in that region. That is the context of the type II bubble move. On the other hand, a birth followed by a death engenders a sphere. A sequence of saddles can change such a sphere into a higher genus surface.

In the discussion of the relationships among $(1 + 1)$-dimensional space times, the idea of a $(1 + 2)$-dimensional space-time is becoming apparent. Here the 2 represents a pair of time-like dimensions. We consider a universe that consists of $(1 + 1)$-dimensional space-times that are interconnected by (oriented) scenes in a movie. The evolution from an empty space-time to an empty space-time is a continuum of possible $(1 + 1)$-universes connected by the singular $(1+1)$-dimensional universes that correspond to the scenes in a movie. The space-time-time continuum of such worlds is an immersed oriented surface in 3-space. The additional requirement that the total number of births and deaths exceeds the number of saddles by 2 ensures that the resulting surface is a sphere.

It is worthwhile to insert a word of caution here. This theory is not string theory as it is understood by physicists! It is, however, a metaphor that oversimplifies the ideas of string theory. The physicists' theories (for there is more than one string theory) includes extra structures on the space-time surfaces, and the interactions that I have described here do not occur in the physicists' theory precisely because those theories happen in higher dimensional spaces.

I hope the next step is obvious. A $(1+3)$-dimensional space-time-time-time is constructed from the time-like surfaces that interpolate between $(1+1)$-dimensional space-time by understanding how these can be related via the movie-moves. The relationships are the strongest possible: they do not include births and deaths of spheres; they do not include handles being attached to the surface; and *most importantly* they do not allow the creation or cancellation of branch points that are only allowed in the context of Chapter 9.

Smale's theorem in this context says that the $(1+2)$-continuum that consists of the birth of a counterclockwise simple closed curve followed by its death is related by a sequence of movie-moves (as articulated here) to the $(1+2)$-dimensional continuum that consists of the birth of a clockwise simple closed curve followed by its death. The connection between these universes is a 3-dimensional solid (time thickened sphere) in $(1+3)$-dimensional space. The relationships between these two movies can be expressed as

$$[(\emptyset) \Rightarrow (\subset^-_+ \supset) \Rightarrow (\emptyset)] \Rrightarrow [(\emptyset) \Rightarrow (\subset^+_- \supset) \Rightarrow (\emptyset)].$$

Thus, *each still in a movie can be given as a symbolic expression.* Each transition between stills can be given as a symbolic expression. And each transition between movies can be given as a symbolic expression. There is little rationale here for doing so, but if one were to ask a machine to manipulate surfaces immersed in space, then one might want to give the machine such codes, and such "rules for rewriting."

Gauss-Morse Codes

C. F. Gauss's life ranges from the last quarter of the 18th century through the first half of the 19th century. His study of knot theory appears in his private notebooks. In his musings, he labels each double point of the projection of a knot with a letter and then records a word in the sequence of letters. In the terminology of this book, the sequence of letters form the double decker set of the immersed curve. An encoding, called *Gauss-Morse codes*, of oriented immersed curves also labels each double point and optimal point. In order to make such a sequence, a direction for the curve is given, a labeling is chosen, and an initial point on the curve is chosen thereby giving the word a starting point and an ending point (unlike the James Joyce novel *Finnegan's Wake*). On the other hand, the sequence of intersection points *does* occur on the circle, and so the Gauss-code (as it is now known) is really a cyclic word. Masahico Saito and I use Gauss codes in conjunction with information about the optima ($\subset$ or $\supset$). So that on a journey around a curve, Maxima, minima, and crossings are encountered and recorded in a word that is the Gauss-Morse code.

Marston Morse hails from the first three quarters of the 20th century. His pioneering study of critical behavior is reflected in every aspect of this book. In the Gauss-Morse code, Maxima are denoted with upper case letters and minima with lower case letters. Thus, Masahico and I pun on the encryption of critical behavior in the sense of Marson Morse and the code of Samuel Morse. Traditional Morse code uses longs and shorts to transmit electronic messages along telegraph wires. The Gauss-Morse code here uses left and right to distinguish critical behaviors. Samuel's three quarters of a century long life encompasses the century prior to the life of Marston.

Since the colors red and blue are associated to the optima, they are recorded as r, or b for minima and R or B for Maxima. For example, an immersed curve and its associated Gauss-Morse code is depicted in Fig. 10.1. The collection of Gauss-Morse codes and their transitions — under each of the scenes that constitute a movie — construct, via interpolation, the decker set of each immersed sphere. Finally, a change between two immersed spheres is encoded by rewriting the sequence of Gauss-Morse codes associated to the movie of each sphere, and interpolating between the two sets of codes via the evolution of the decker set as in Chapter 7.

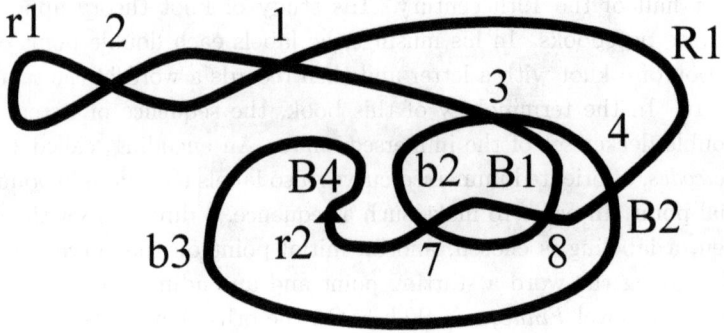

$$B4,b3,B2,4,3,b2,7,8,B1,3,1,2,r1,2,1,R1,4,8,7,r2$$

Fig. 10.1 An immersed curve and its Gauss-Morse code starting from the arrow

Summary

Starting from a simple-minded physical metaphor, the complications of the sphere eversion can be constructed as the relationships among the relationships among 1-dimensional universes with balanced charges and an even number of particle exchanges. This schema allows the eversion to be codified in an algebraic fashion. An alternate codification is used to construct the decker sets and their evolution.

Chapter 11

Sarah's Thesis

Sarah Gelsinger Brewer's master's degree thesis consists of the explicit computation of the topological type of the fold set, the double point set, the double decker set, and the cusp set of the immersion depicted here. The original collection of ink and colored pencil drawings of the sketches for this book spanned a decade's worth of computations. Steps are skipped. The computations sometimes proceed from red to blue, sometimes from blue to red. The pages might be out of order.

A student learns to read her advisor's mind. Sarah's contribution here includes draft drawings of the decker sets and the movies, the corrections to a number of technical points both in the drawings and in the mathematics, and the calculations that are described in the current chapter.

In this chapter, the details of Sarah's thesis are presented and the notion of topological type is summarized.

———

A string that evolves in time sweeps out a surface. The critical events in the process are births and deaths of loops and the reconnection between arcs affected by a saddle. Such a surface is an abstract quantity independent of its relationship with the surrounding environment. For example, the double point surface of the eversion is an abstract surface that also happens to contain the self-intersections of the evolving sphere. Similarly, the fold surface is an abstract surface with seams formed by the evolving cusp set. In general, an abstract surface *can* be viewed as the evolution of arcs and loops; sometimes this viewpoint is useful and sometimes the surface needs to be seen as an entire entity.

———————

A magnificent theorem of topology allows any surface to be identified by looking at the number boundary components, the number of births and deaths, the number of saddles, and whether or not the surface carries a global notion of left and right. This classification theorem was used by Sarah to determine the nature of the double point surface and the fold surface. Sarah's computation indicates that the double point surface is *non-orientable* in the sense that left and right cannot be chosen in a global sense. In fact, *any* eversion has a non-orientable double point surface. The classification theorem allows one to see *which* non-orientable surface the double point set forms.

The archetypical example of surface without a global notion of left and right is the Möbius band. Every non-orientable surface contains a Möbius band within it. Therefore, the Möbius band is the most important example in topology. It is also one of the most written about examples, and even though you may be familiar with many of its miraculous properties, there is one that is essential to the current discussion: a saddle between closed curves that does not change the number of components is a half-twisted band, and the interpolating surface therefore contains a Möbius band. Figure 11.1 illustrates. Each stage in movie of any surface may be oriented, but some saddles make it impossible to have a consistent orientation from one still to the next. The figure is drawn with the saddle occurring between the left and right curves, and the surface is predominantly shaded purple since such saddles occur on the double point set.

So, while the double point set could be oriented at each stage of the eversion, there are some type II saddle moves which do not change the number of circles in the double point set, and consequently the orientation on the double points cannot be chosen consistently throughout the eversion. As the sphere morphs from red to blue, the points at which non-orientable saddles occur on the double point set are easily identified.

In Sarah's thesis, the double point set of this eversion is computed to be non-orientable and to have two births, two deaths, and five saddles. The classification theorem of surfaces allows us to determine that the surface can be identified with a sphere that has had three disks removed and these have been replaced by three Möbius bands. Such a surface can never be put into 3-dimensional space without passing through itself. But as the double point set, it does pass through itself since there are triple points upon it.

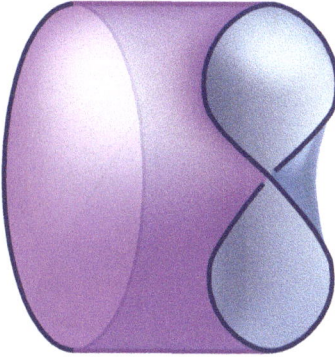

Fig. 11.1 A Möbius band results when a saddle does not change the number of components

Its schematic view is illustrated in Fig. 11.2.

———————

The fold set of the eversion starts from a red circle of folds and ends at a blue circle of folds. The fold set is an orientable surface that has a red and a blue circle forming its boundary. The fold set intersects itself in space-time at each horizontal type II move. On the time-elapsed sphere it is embedded. By counting all of the critical steps in the evolution of the fold set, Sarah determined that the fold set has the topological type of an annulus or gasket. The time-elapsed view of the cusp set forms a simple closed curve on this annulus. This is a seam upon which the red fold surface and the blue fold surface are sewn together. The time-elapsed view of the births, deaths, and saddles form a pair of arcs on the fold surface. Interestingly, these arcs change colors at the point of a horizontal cusp.

A tool that aids Sarah's compuation is the illustration in Fig. 13.4. which is included in Chapter 13. It represents the time-elasped summary

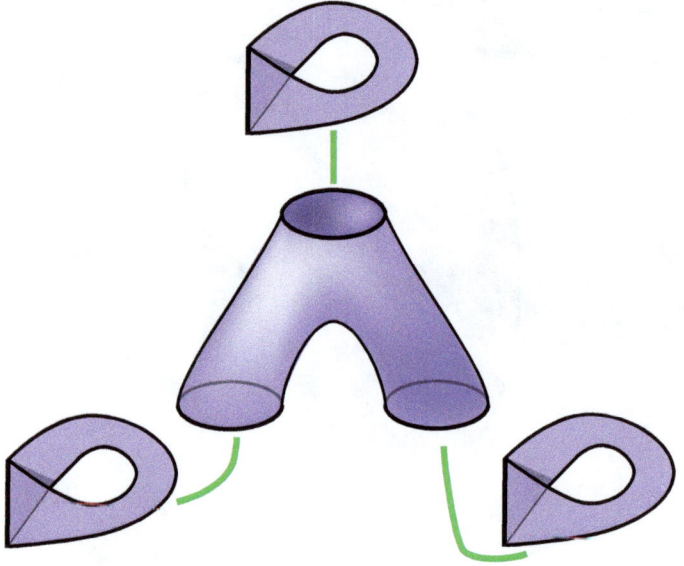

Fig. 11.2   The double point set as an abstract manifold

of all of the critical events in the eversion process. Within the illustration, the optimal points, the saddle points, the cusps, the optima of the double points, and the triple points are tracked. The figure illustrates the profile of the process, and it aids us in envisioning the eversion process as a single space-time entity.

Chapter 12

The Eversion

In my lectures about this eversion, two senior colleagues, for whom I have the greatest respect, ask me the same questions, "What is the guiding principle in this eversion? How do you know you are making progress? How can you explain what you are doing from a global perspective?"

The theorem of Banchoff and Max states that every sphere eversion has a quadruple point. The singular point of Outside-In, for example, has multiplicity much greater than 4. And Outside-In, although elegant, is highly singular and consequently fine details are difficult to discern. Starting from the observation that there must be a quadruple point, the *red side of the eversion* is defined to be the sequence of steps before the quadruple point occurs.

In order to create a generic quadruple point, there must be four triple points interconnected by six arcs of double points such that this configuration forms a tetrahedron. Four embedded triangles on the sphere form the tetrahedron. So the goal on the red side is to create two pairs of triple points and arrange the double point set so that these four triple points form the boundary of a tetrahedron. This goal was achieved by watching an animation of the Froisart-Morin eversion and trying a minor variation of this.

To get to the blue side, pass the triple points through the tetrahedron. After the quadruple point, one can recognize that a pair of triple points can be easily canceled via a type III-type III move. Now at that stage, it is easy to be stuck for about two weeks because the triple points that should cancel are found on different sides of the sphere and it is not clear how to bring them close enough in a movie description. Much like a geometric puzzle that is called a *quebra de cabeça* in Portuguese, I look at this stage from a variety of points of view before adding a pair of cusps connected by lips. The folds at the lips make the surface flexible enough to move the triple points near each other.

In retrospect, the introduction of lips is clearly necessary. The fold set of the blue sphere is a blue circle. One of the folds in the introduced lips eventually evolves into this fold circle. A subsequent creation of a maximum-saddle pair allows the triple point that is towards the back side of the movies, to move forward. Within the diagrams, one can see the top portion of the sphere apparently twisting. That twisting action brings the triple points close together so that they can cancel. The rest of the process becomes quite simple.

During the twisting portion on the blue side, I think that we are witnessing Thurston's belt trick occurring. Smale's original proof requires some algebraic computations. The fact that a doubly twisted belt can be undone in the space of rotations of space is an ingredient in Smale's proof. So I expect that one can see a belt trick in every eversion. Others whom I know also believe this is true. The beginning mathematician who is reading this may wish to prove the general statement.

The text now turns to illustrating each step in the eversion. Numbers $1, 2, 3$, *etc.* indicate arcs of double points. A lower case r indicates a red fold on the left, the upper case R indicates a red fold on the right, a lower case b indicates a blue fold on the left, and B indicates a blue fold on the right. Indices are carried forward from still to still and from picture to picture. Even numbered pages contain the movie and projection of the sphere onto the plane. The left illustration indicates a semi-transparent view, the center picture indicates the folds and double points, and the right picture illustrates the surface as it would appear with all surfaces opaque. Line thickness and degrees of transparency indicate the number of veils of surface between you and the segment in question. Surfaces below the first depth are shaded with the color that faces you. But not every surface facing you is colored. The coloration is chosen to suggest the surface underneath, but sometime colors are removed for clarity. I do not have a consistent degree of transparency and choice of layer that conveyed the surfaces as I conceived them. All the information can be obtained from the associated movie.

The indices of the double points and the fold arcs are inherited from the original hand-drawn illustrations. These are indicated on the top of the figure. The number choices 1,2, *etc.* of double point arc labels were made at the time of the illustrations thinking of the movie starting at the top. Thus 1 and 2 die together and interactions and changes occur earlier in time. As the eversion progresses sequential numberings of arcs that are born together rarely occur.

On the odd numbered pages the decker sets are illustrated. The indexes on the decker here coincide with those in the movies and on the surfaces. Each arc of double points forms a pair of arcs on the double decker surface.

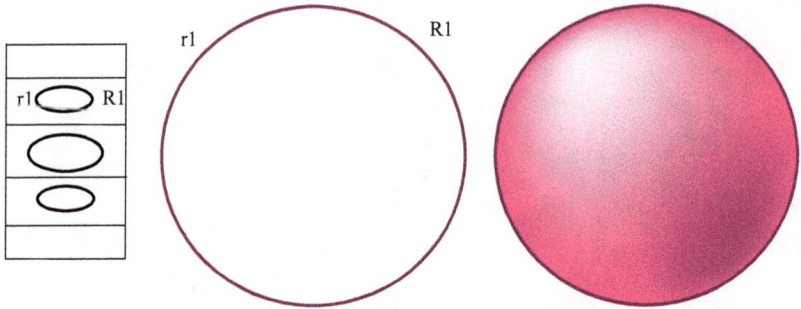

The eversion starts at a red sphere. The next stage is the introduction of a pair of red lips with the visible fold on the left.

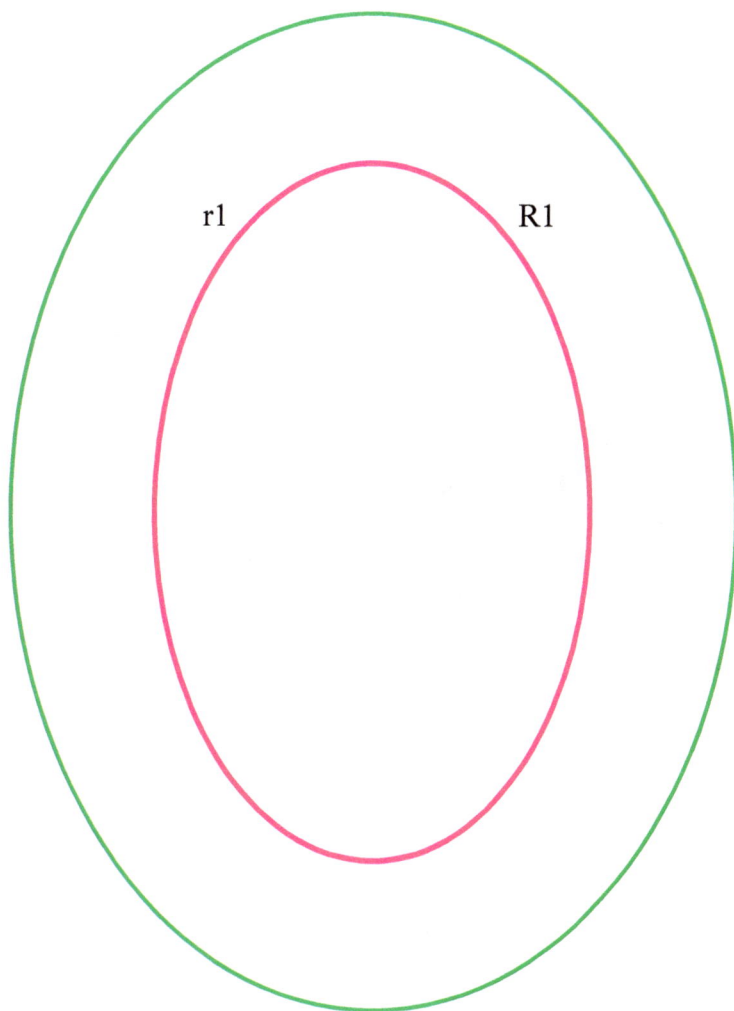

r1   R1

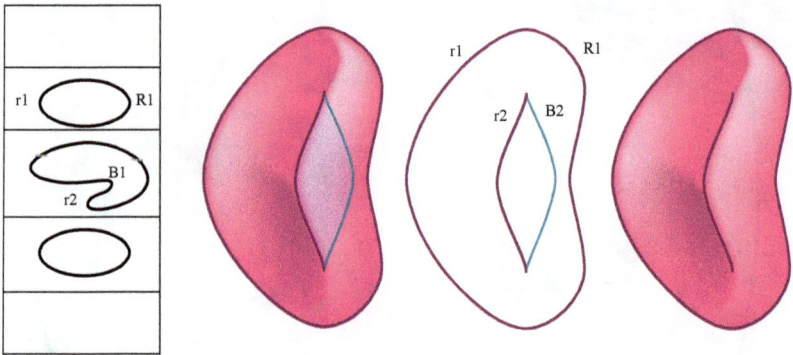

This is illustration red # 2. The next stage is the introduction of a loop of double points via a type II bubble move.

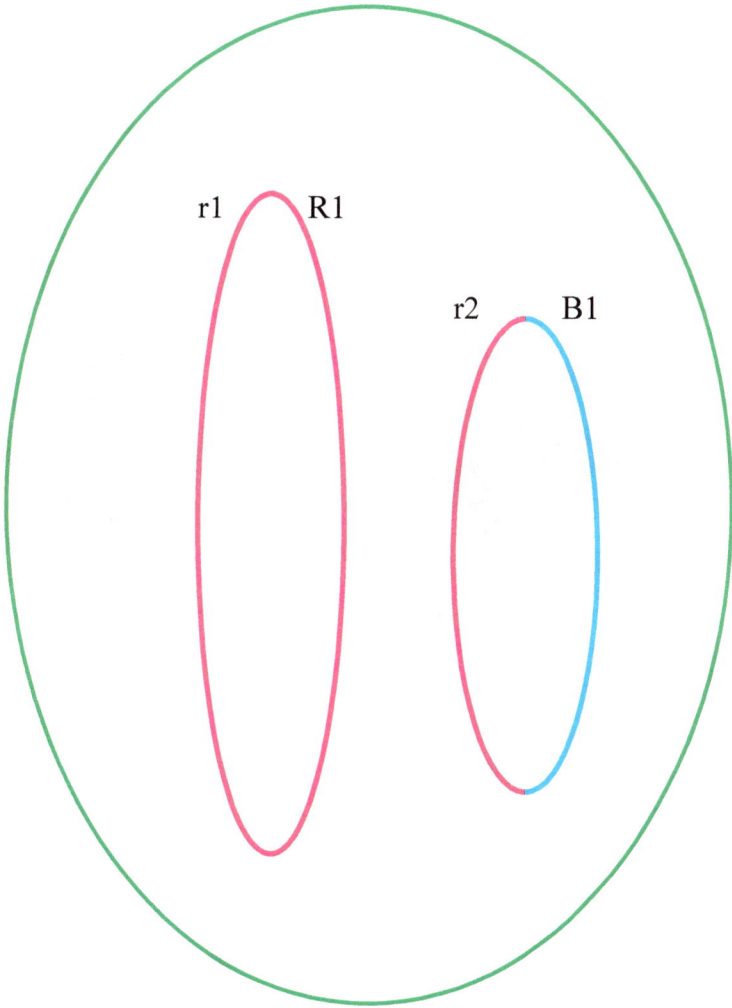

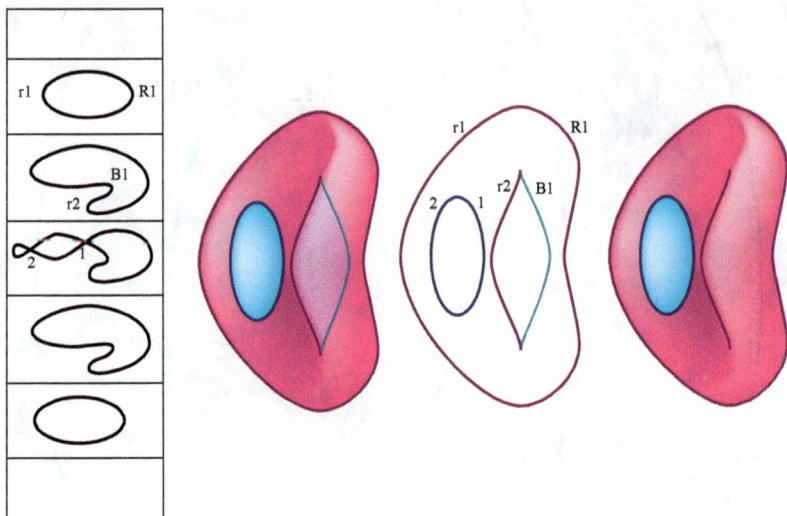

This is illustration red # 3. In the next stage, there is a critical exchange between the double point arc labeled 1 and the red fold $r2$.

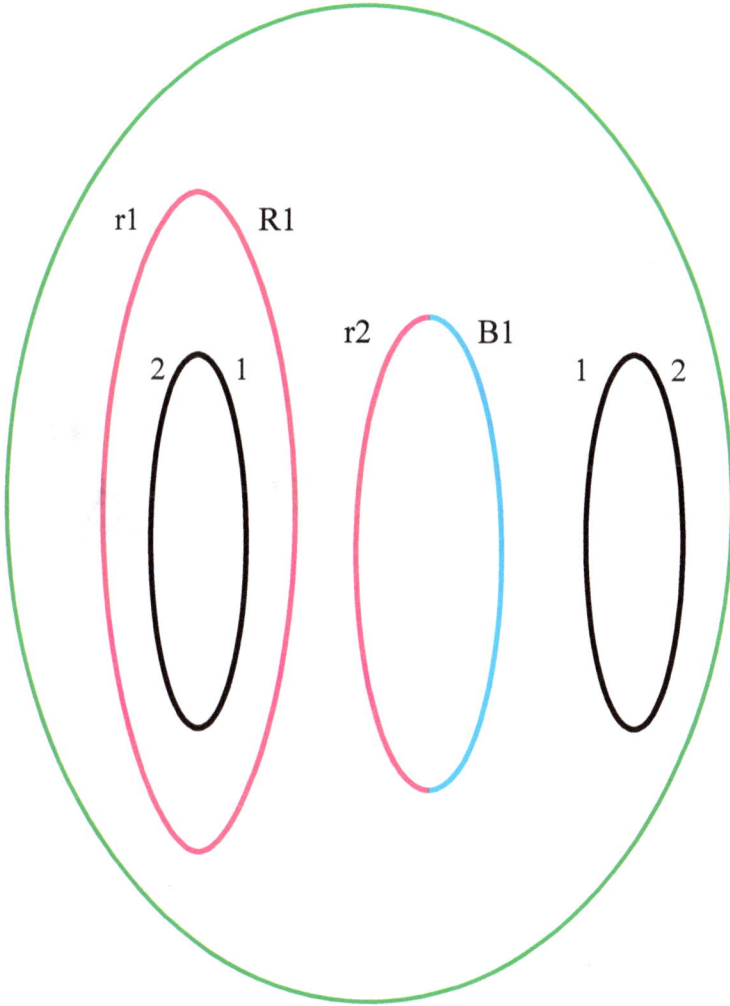

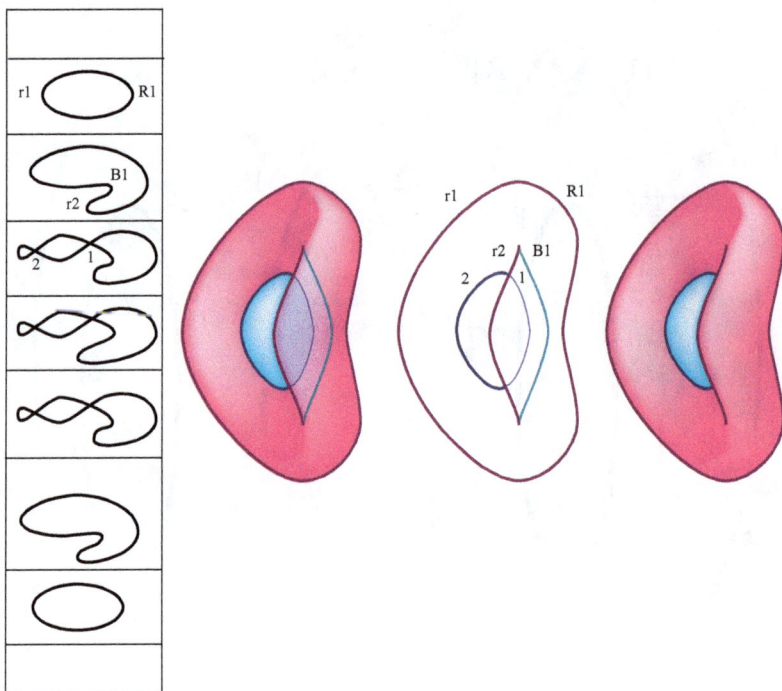

This is illustration red # 4. In the next stage, there is a ψ, ψ-move between double point arc 1 and blue fold $B1$

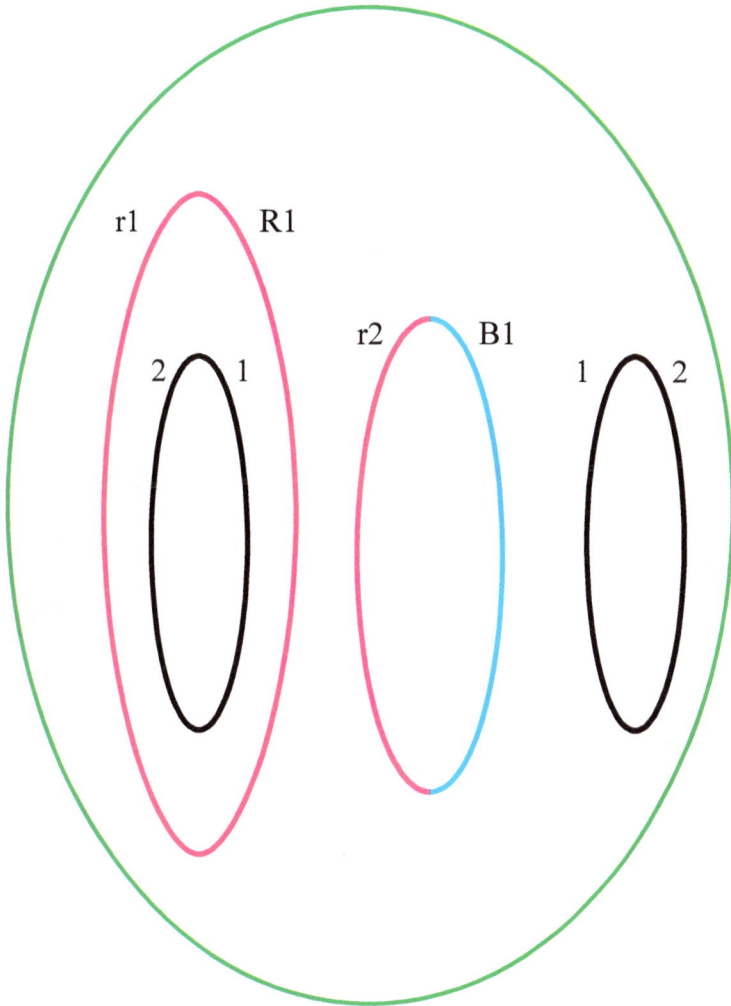

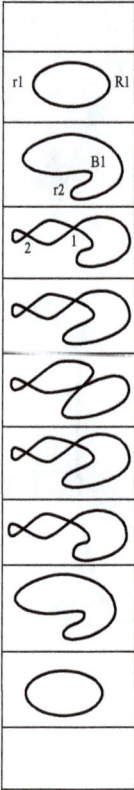

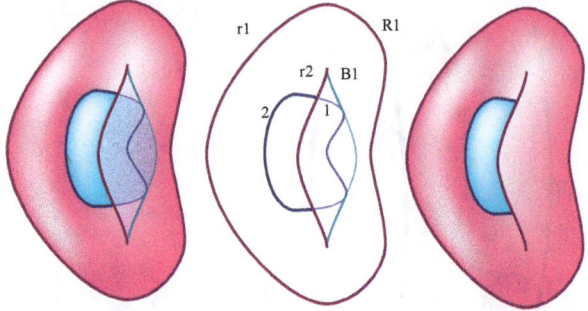

This is illustration red # 5. In the next stage, the blue fold $B1$ peeks out on the right of $R1$ by a canceling pair of critical exchanges.

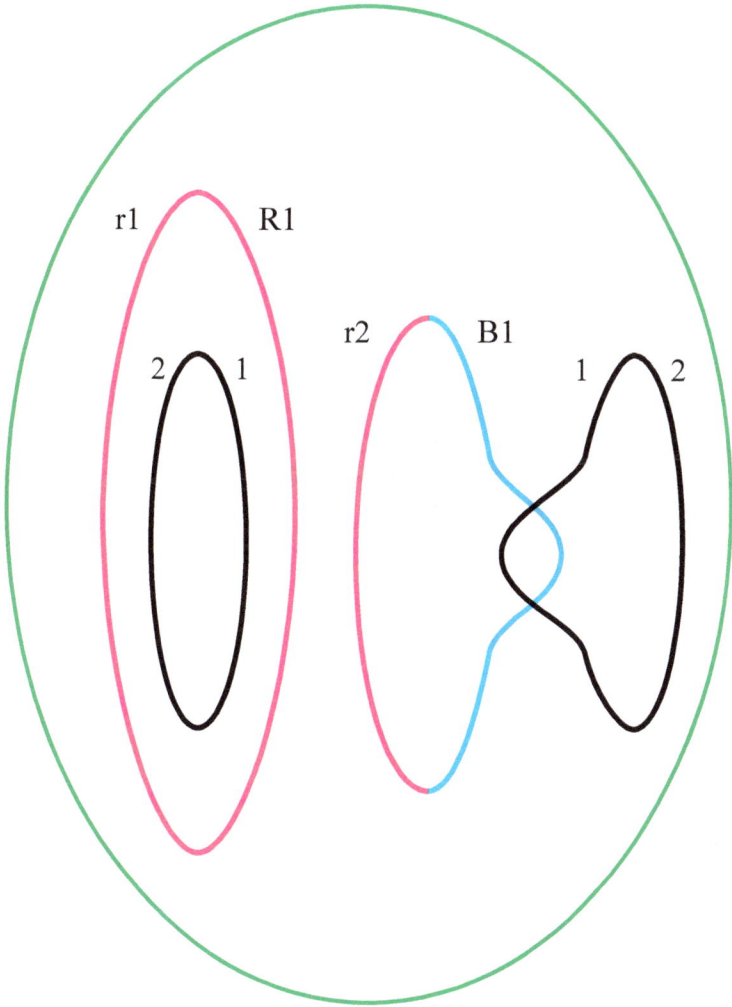

This is illustration red # 6. In the next stage, a type II bubble move occurs on the right introducing a loop of double points with labels 3 and 4.

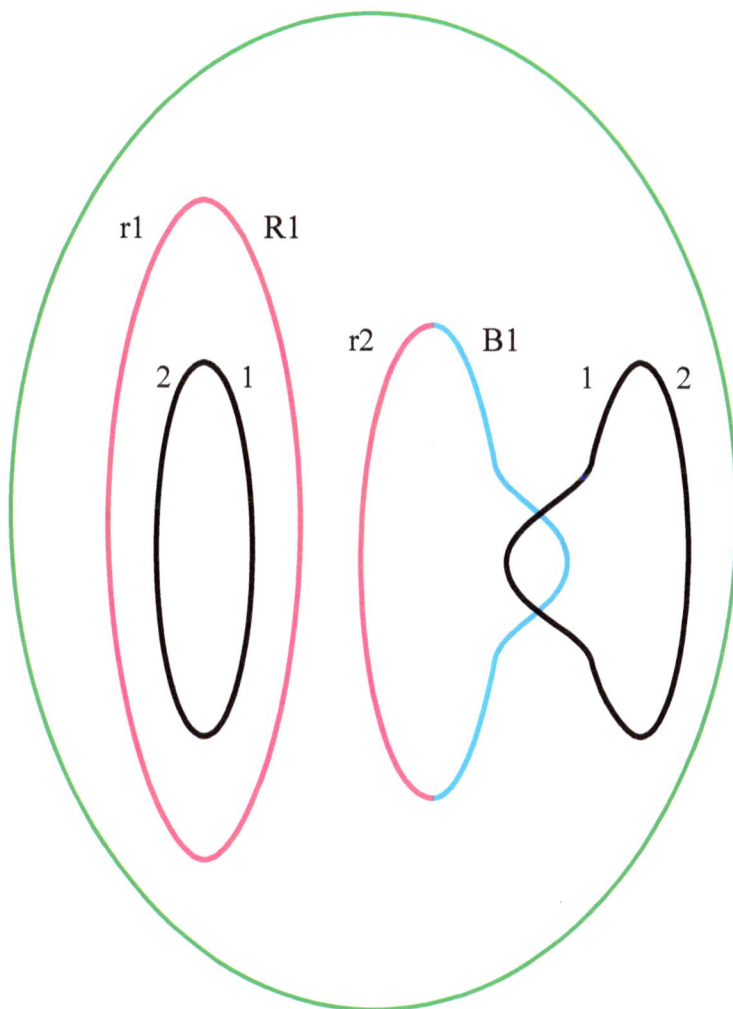

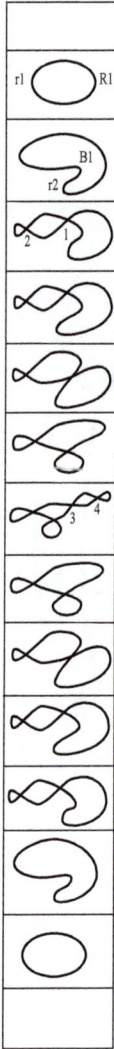

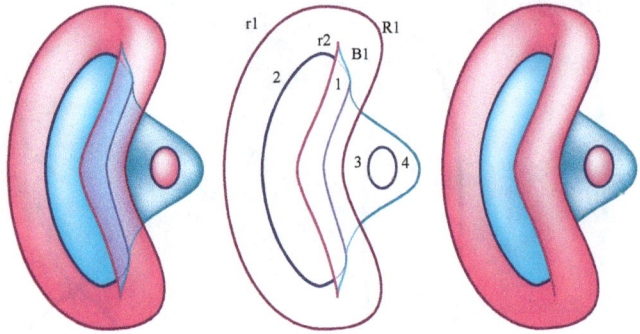

This is illustration red # 7. In the next stage, the double point arc 3 hides behind fold $R1$, arc 2 hides behind fold $r2$ via two pairs of critical exchanges; immediately thereafter a pair of triple points will be introduced by a type III-type III move.

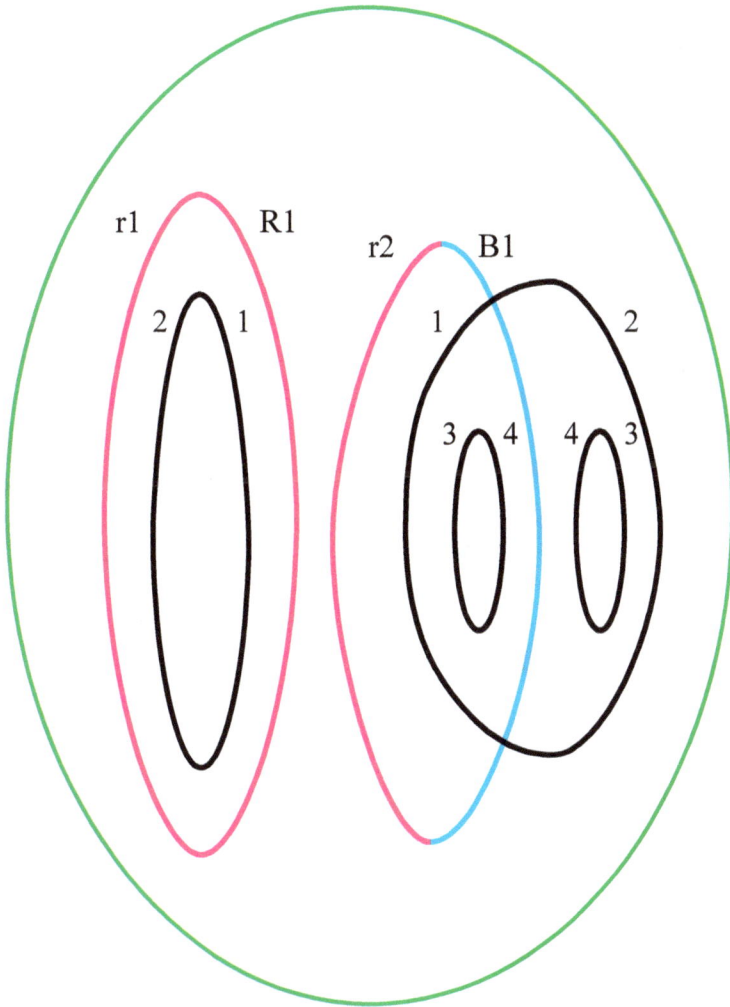

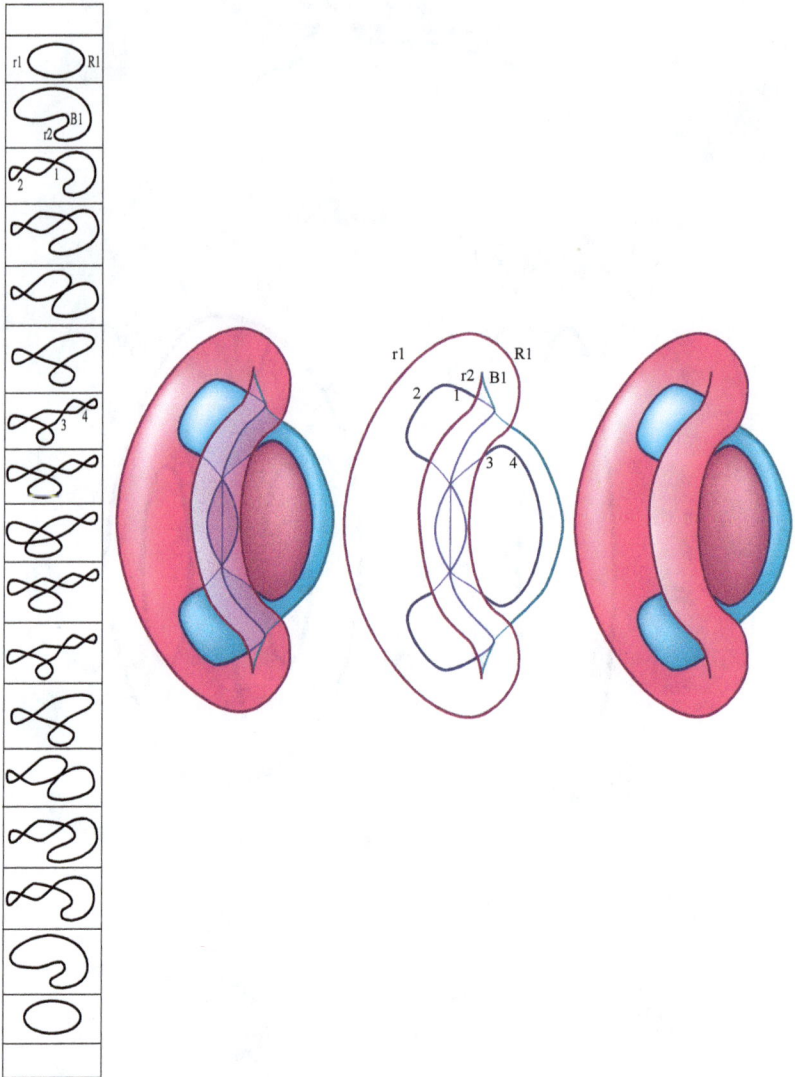

This is illustration red # 8. In the next stage, the double point arc 2 will bounce to the front of $R1$ via a ψ, ψ-move that occurs in the region between the two triple points. The fold $r1$ moves behind the fold $r2$ by a critical exchange.

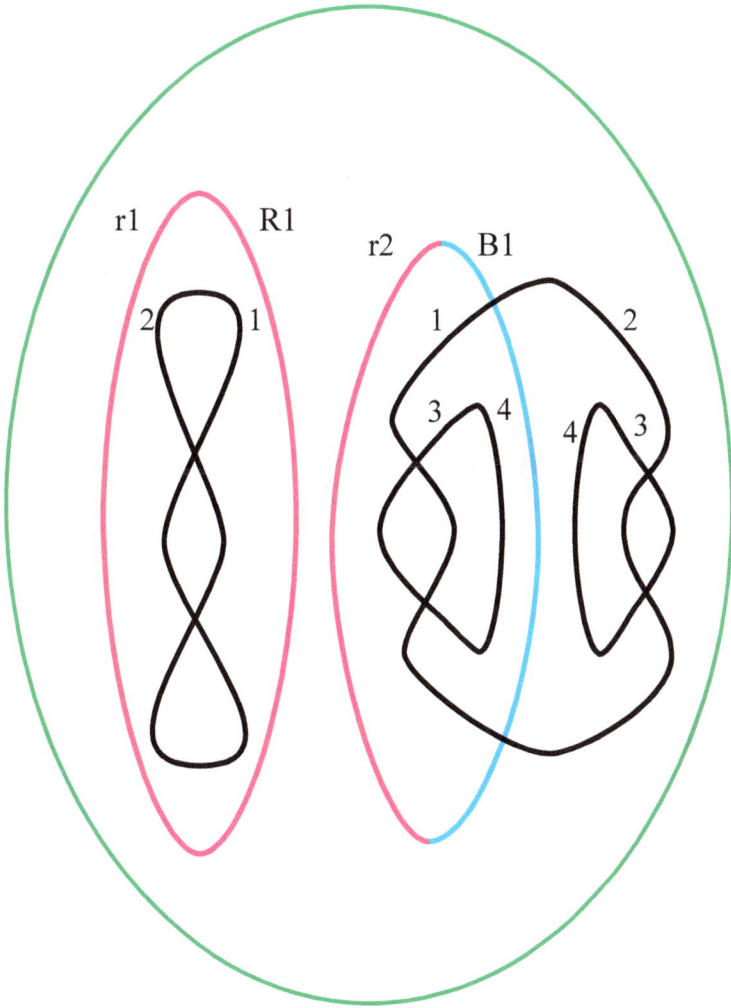

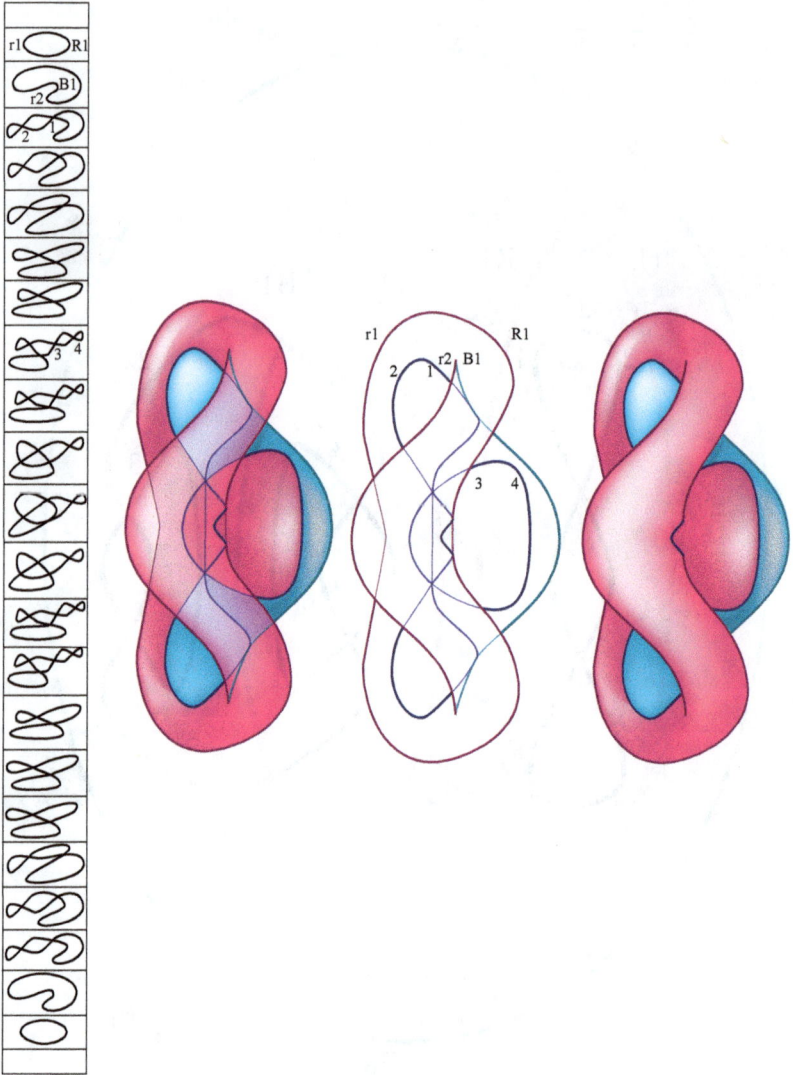

This is illustration red # 9. In the next stage, in the region between triple points, the double point arc 1 exchanges horizontal positions with the double point arc 2 and then bounces to the front of $R1$ via a ψ, ψ-move. In this way, an arc of $R1$ will lie behind the rest of the immersed sphere.

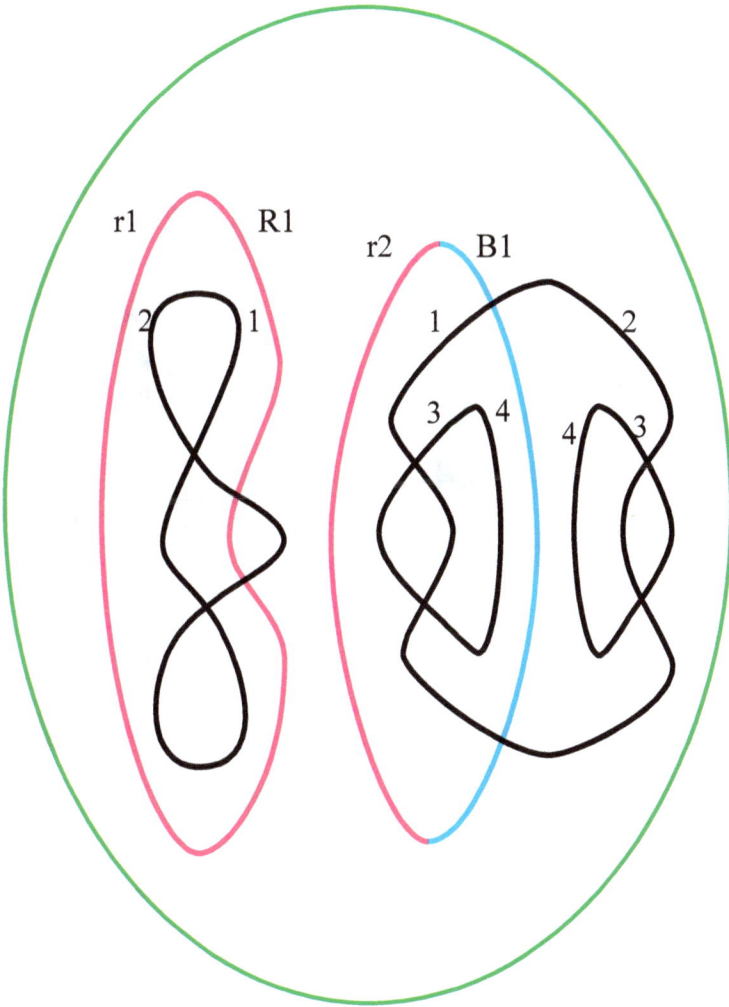

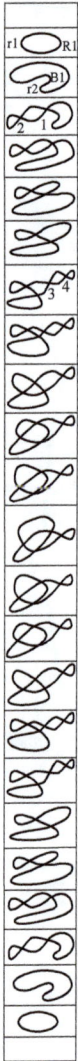

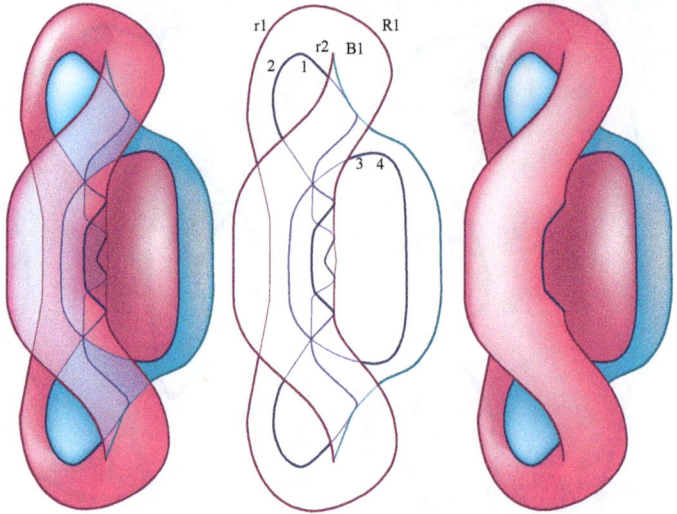

This is illustration red # 10. In the next stage, the double point arc 1 will bounce to the front of $R1$ at the bottom via a ψ, ψ-move at the bottom of the illustration. Observe that within the movie, the migration of fold $r2$ towards the left occurs simultaneously with some other events.

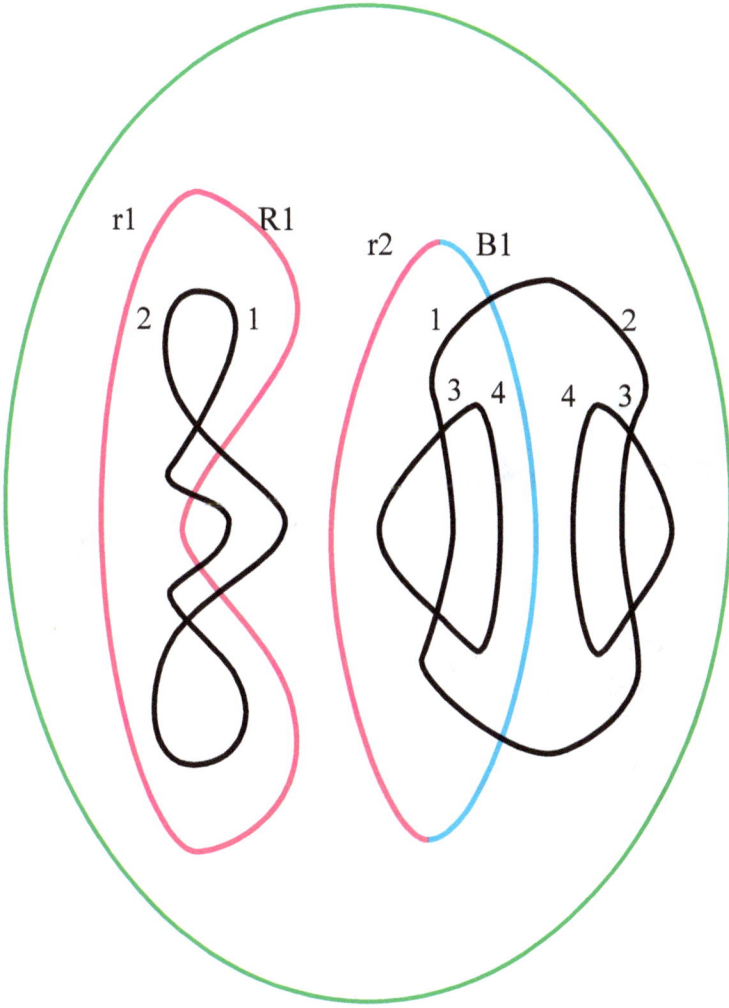

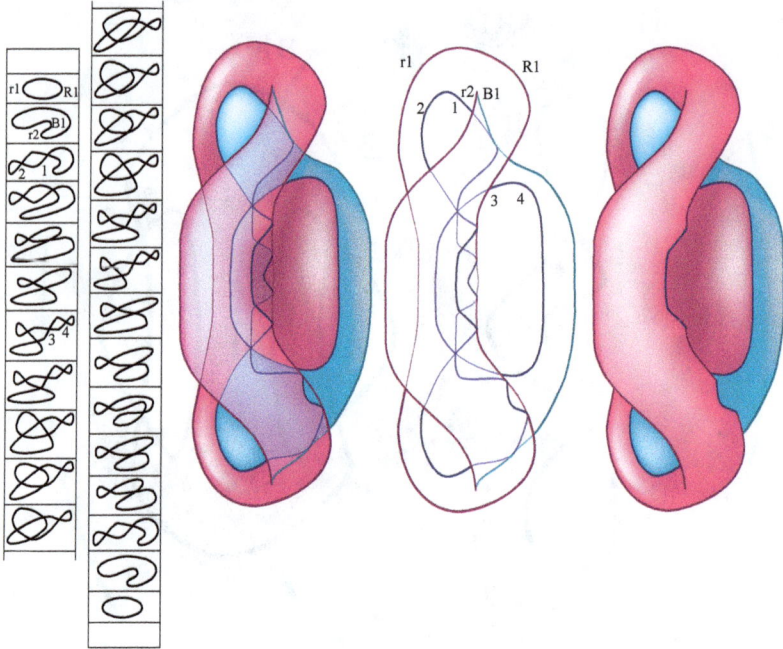

This is illustration red # 11. In the next stage, the double point arc 2 moves to the right of arc 1 at the bottom of the figure and then also bounce to the front of $R1$ via a ψ, ψ-move at the bottom of the illustration.

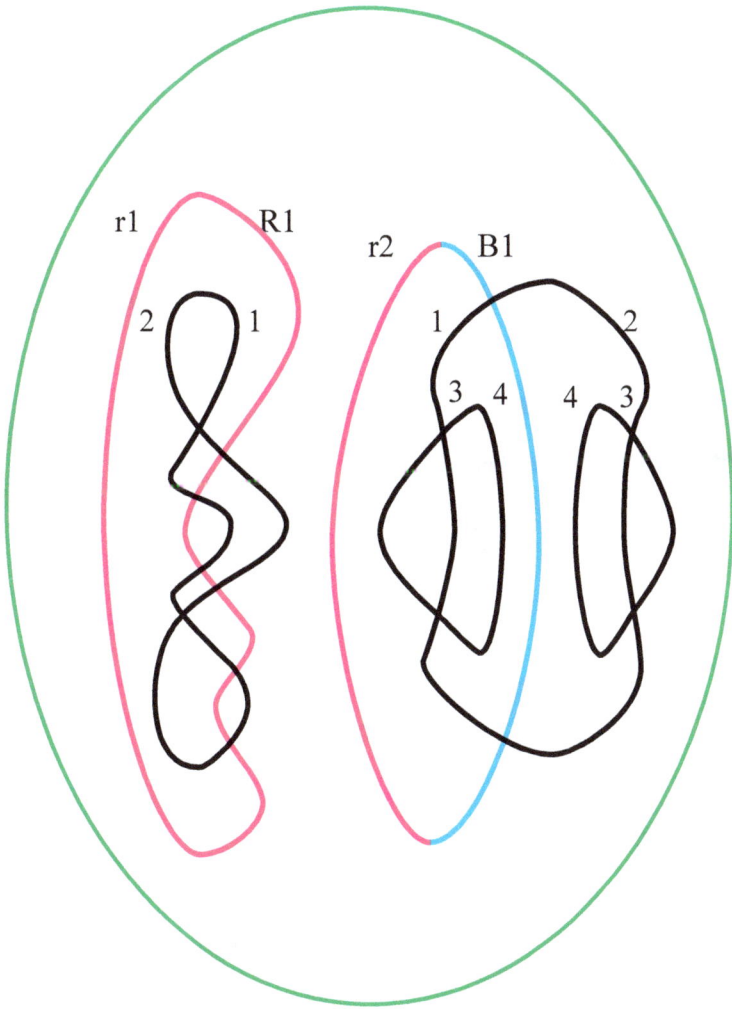

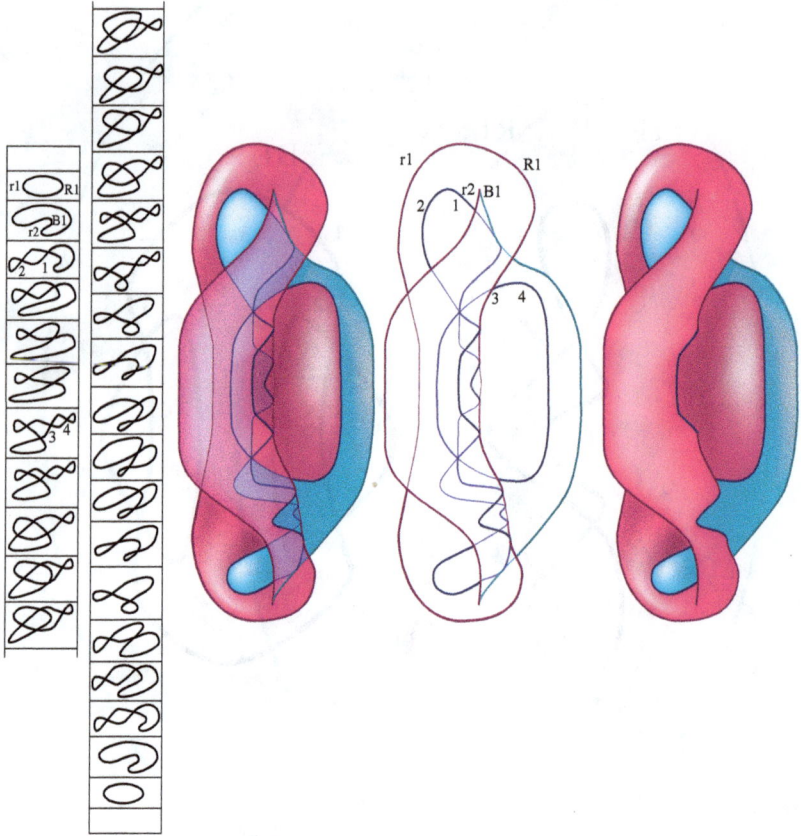

This is illustration red # 12. In the next stage, the triple point at the bottom passes to the front of fold $R1$, fold $r1$ no longer exchanges places with $r2$, and redundant bounces on the bottom are removed. The birth of double points 1 and 2 at the bottom move earlier than the cusp between $r2$ and $B1$.

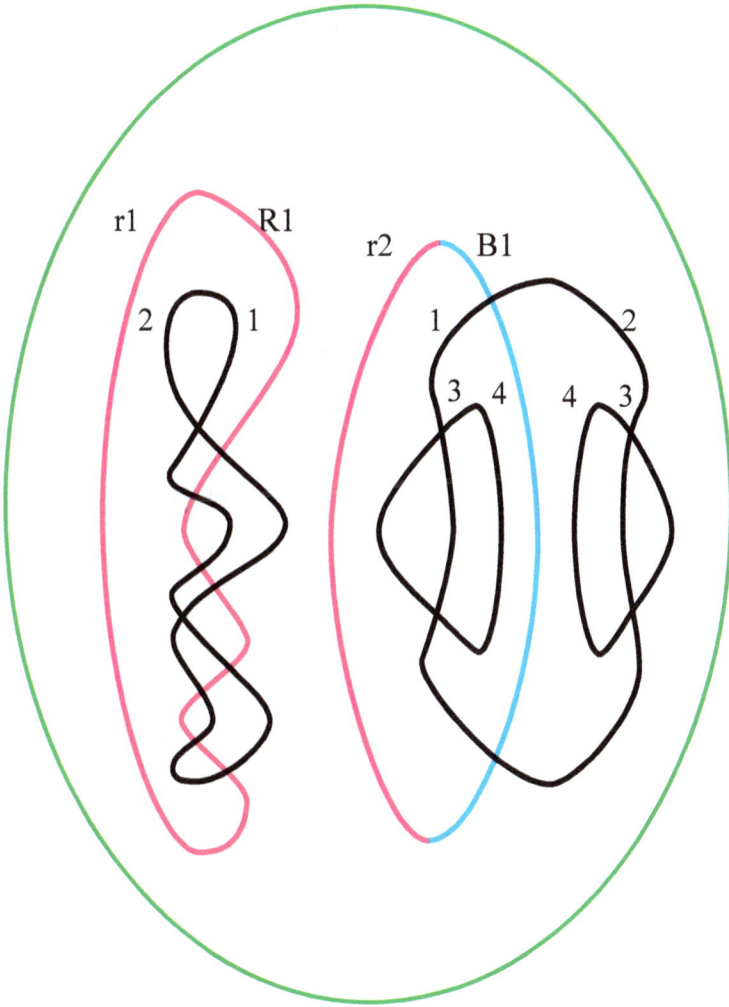

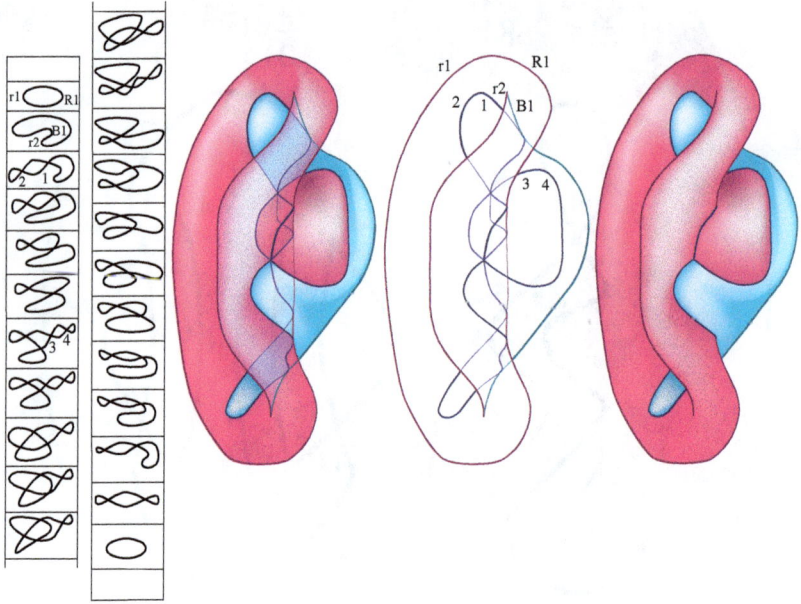

This is illustration red # 13. In the next stage, a horizontal type II move occurs at the bottom, the birth of the double points 1 and 2 hides behind the fold $r2$ and ψ, ψ-moves are performed to remove redundant bounces.

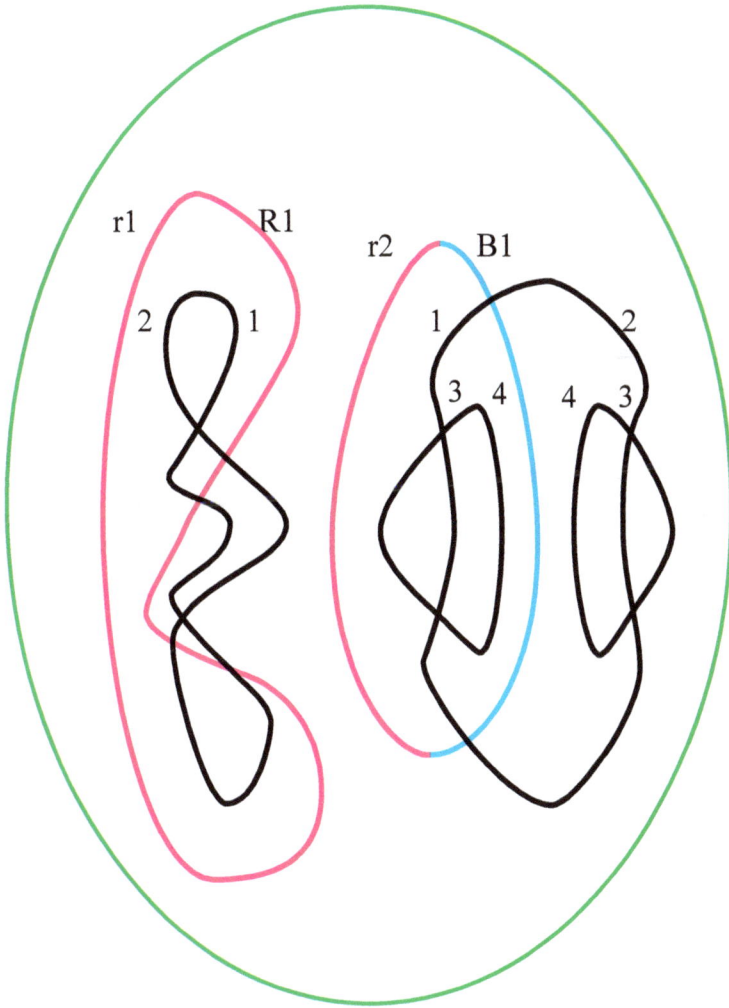

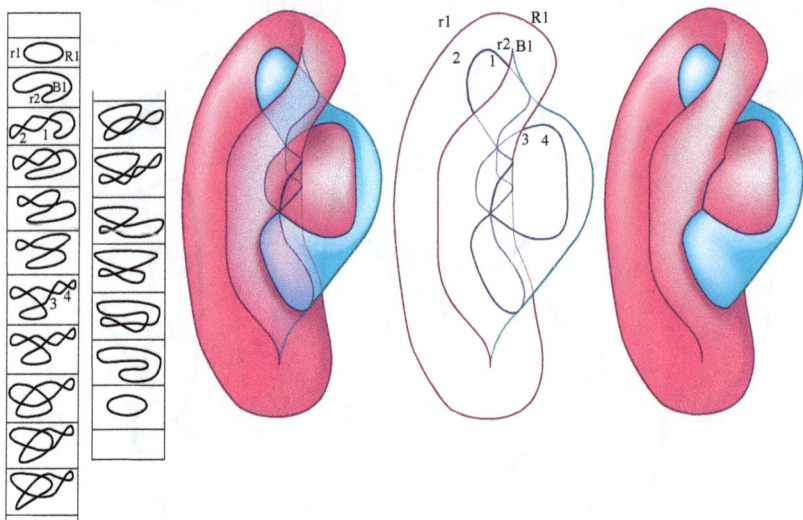

This is illustration red # 14. In the next stage, fold $r2$ crosses back to the left of $r1$, a ψ, ψ-move occurs between fold $r1$ and double point arc 3, and a type II saddle occurs on arcs 1 and 3. The sequence of moves is possible since the bounces on fold $r1$ can interchange positions with those on fold $B1$.

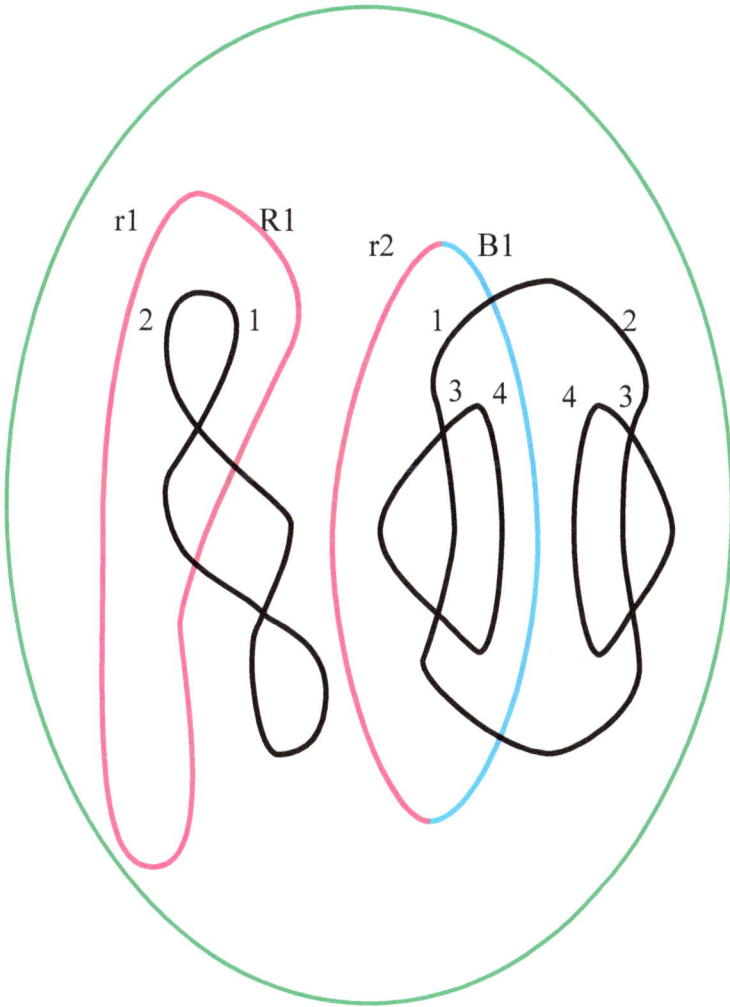

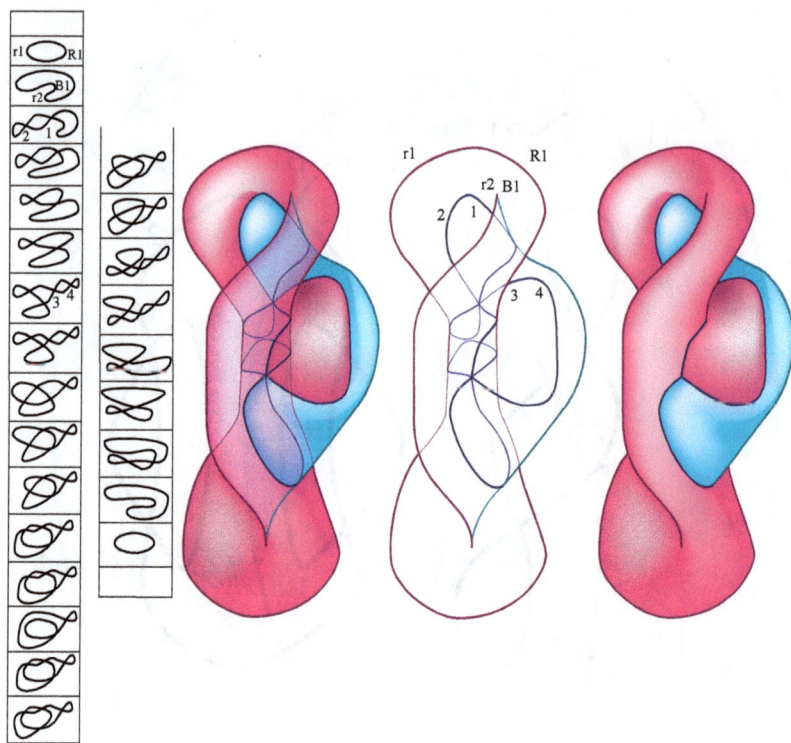

This is illustration red # 15. In the next stage, the bottom triple point moves to the inside of fold $r1$.

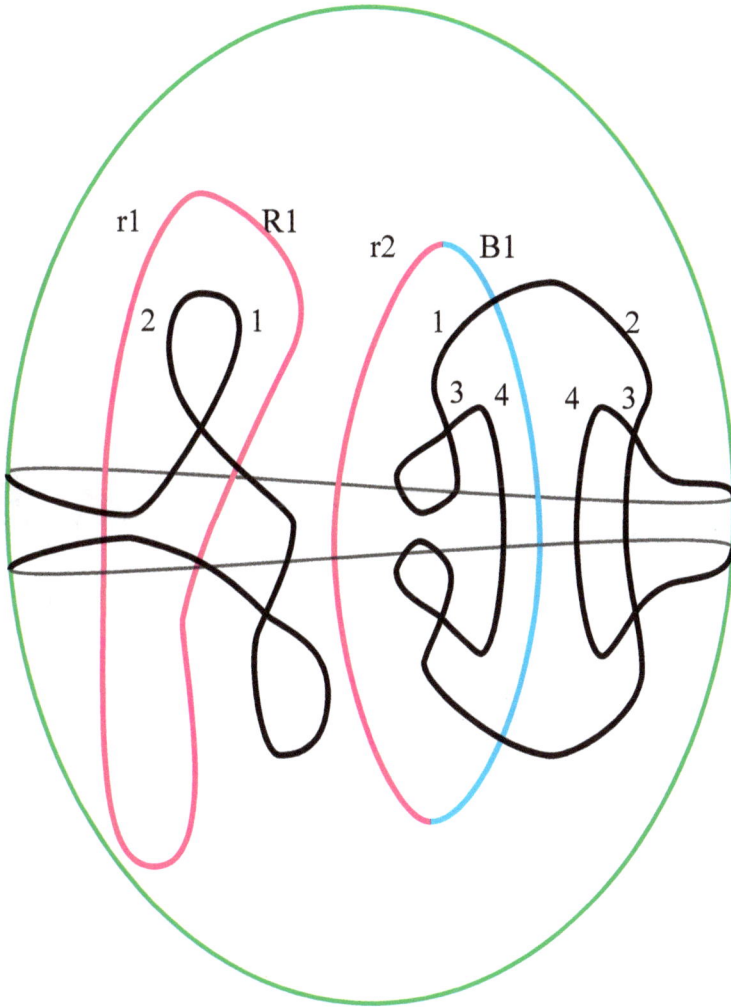

Notice on the decker set here, the arcs connecting 1 and 3 pass to the back of the sphere.

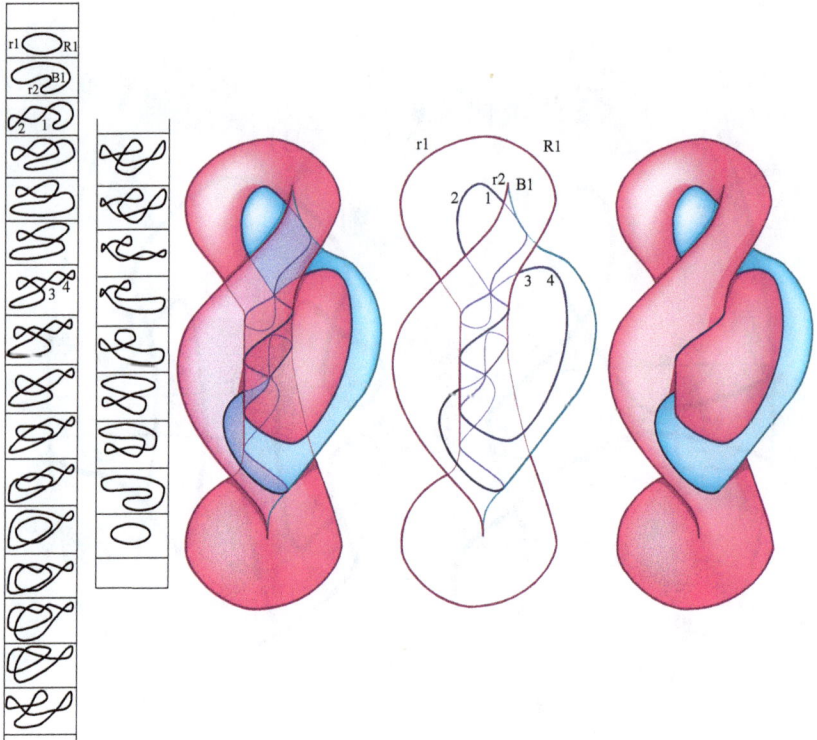

This is illustration red # 16. In the next stage, a beak-to-beak move occurs at the bottom between folds $r1$ and $B1$.

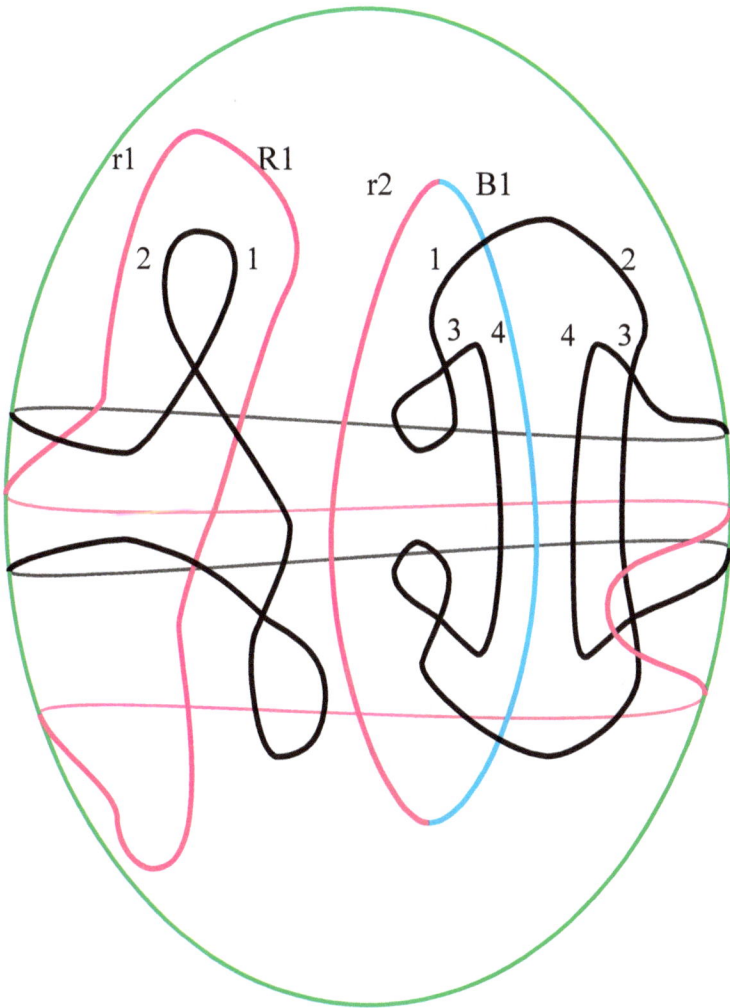

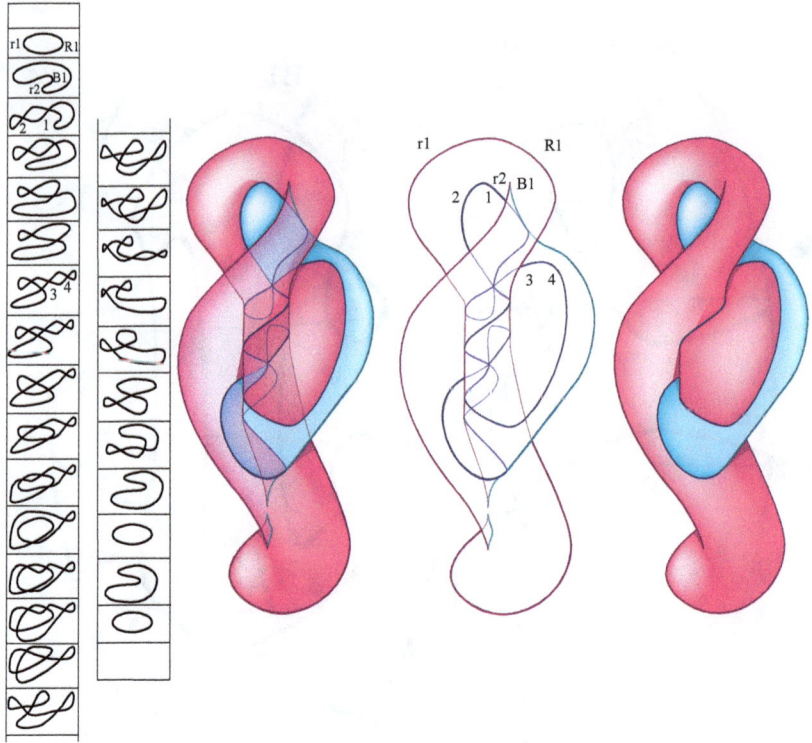

This is illustration red # 17. In the next stage, a swallow-tail move occurs at the bottom.

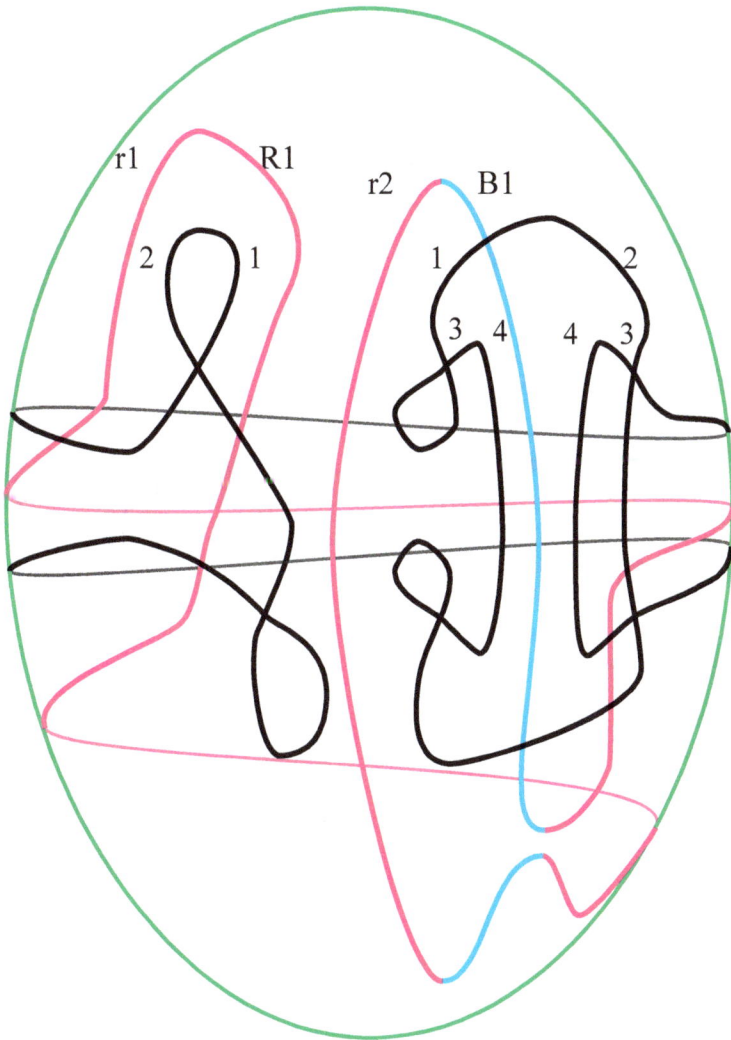

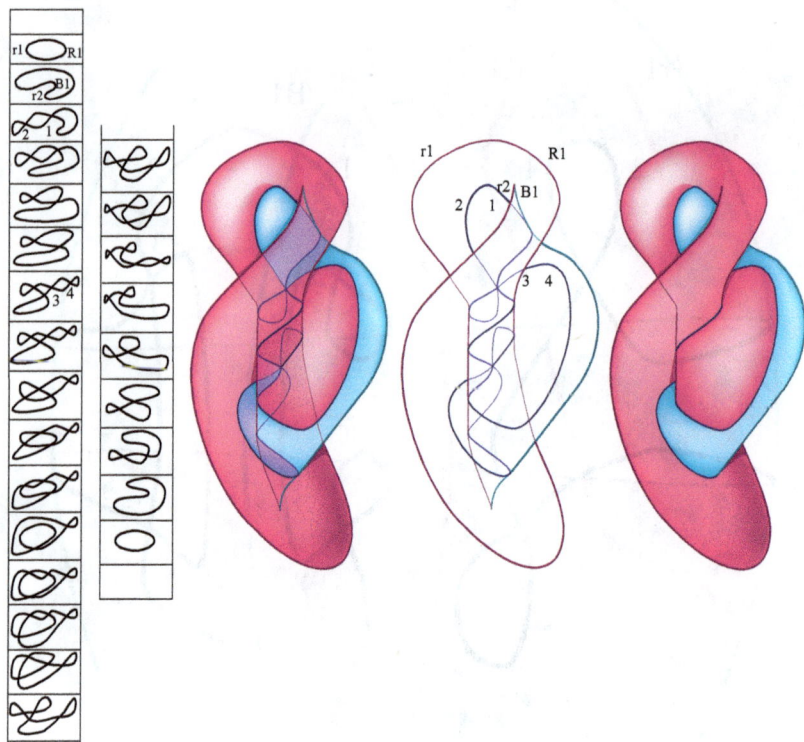

This is illustration red # 18. In the next stage, the double point arc labeled 2 bounces to the other side of the bottom most cusp.

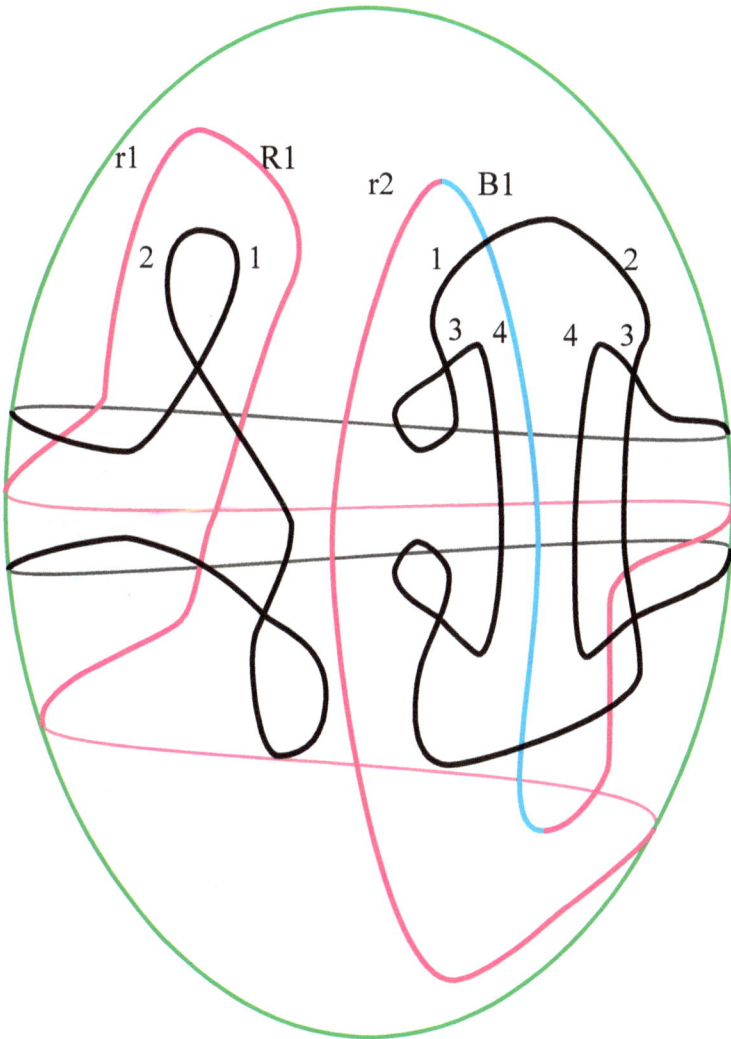

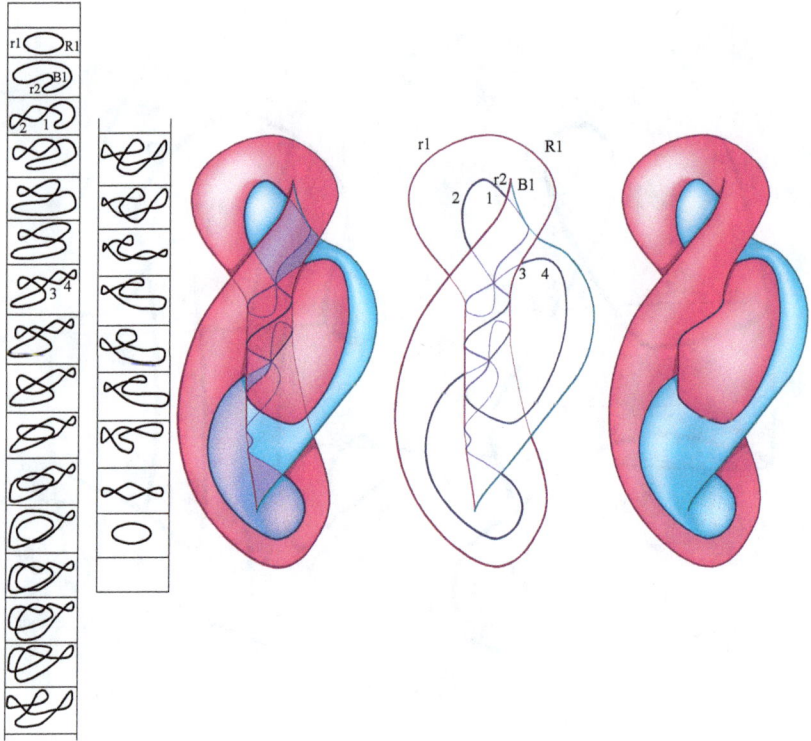

This is illustration red # 19. In the next stage, a ψ, ψ-move removes the redundant bounce between double point arc 2 and the fold $r1$.

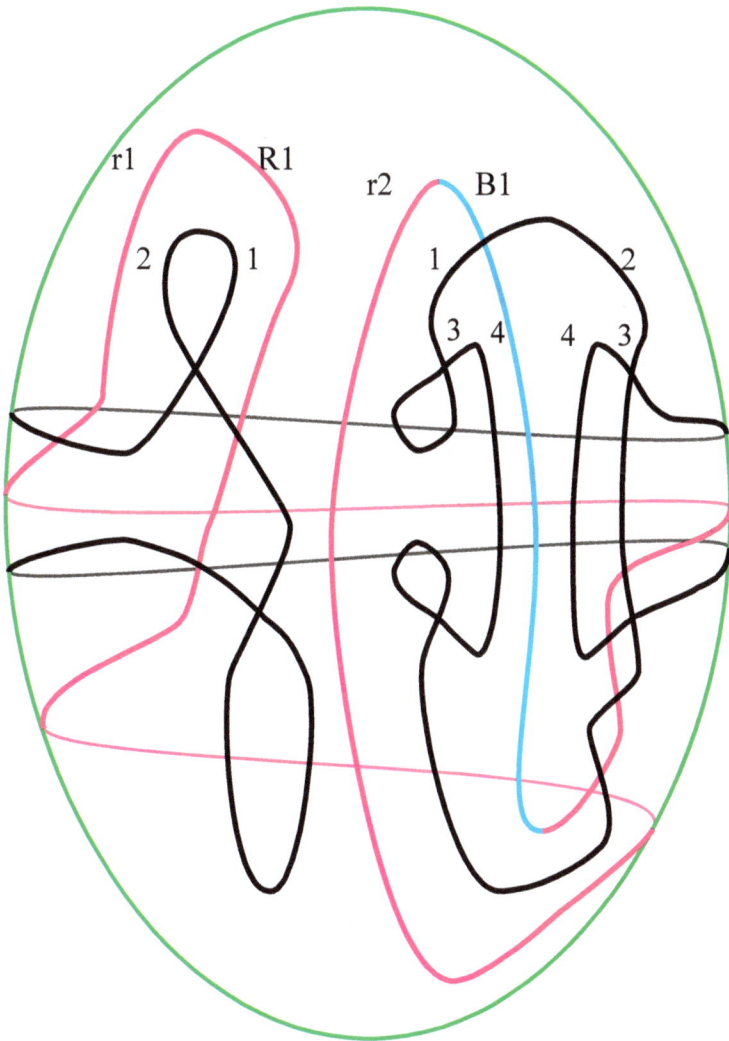

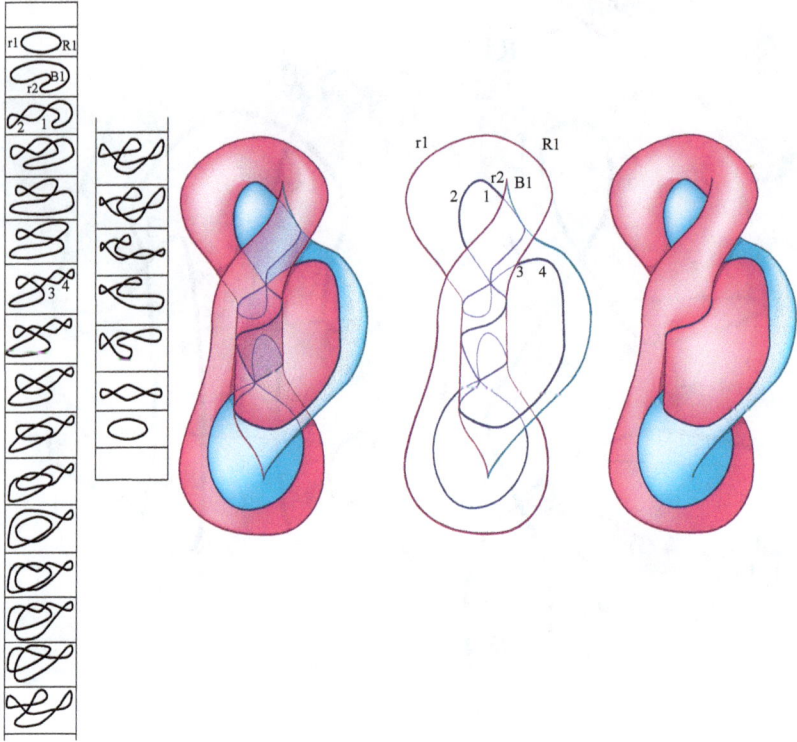

This is illustration red # 20. In the next stage, a ψ, ψ-move occurs below the top triple point. The double point arc 3 bounces in front of the fold $r2$ and a subsequent horizontal type II move occurs at the same place.

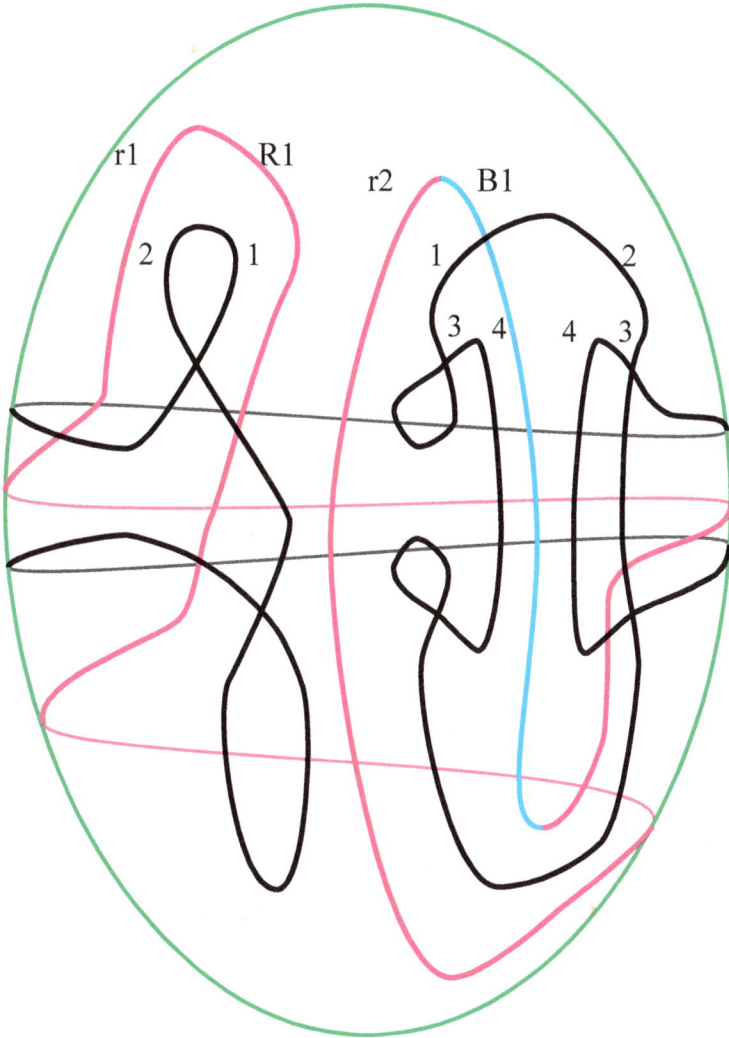

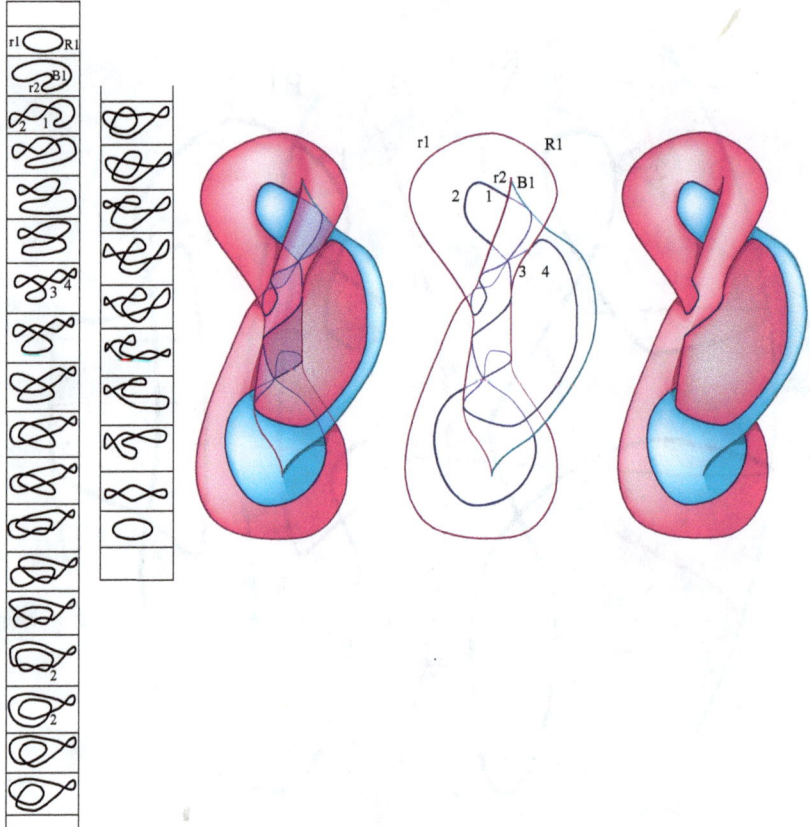

This is illustration red # 21. In the next stage, a type II saddle move occurs between the arcs 3 and 2 between the triple points.

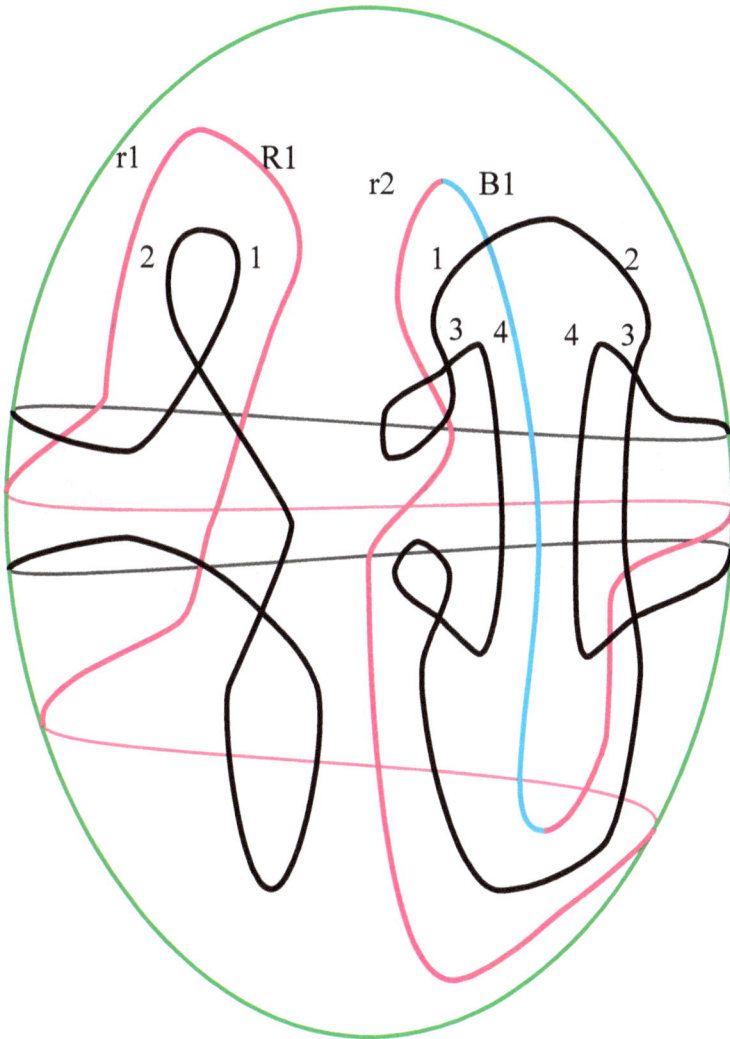

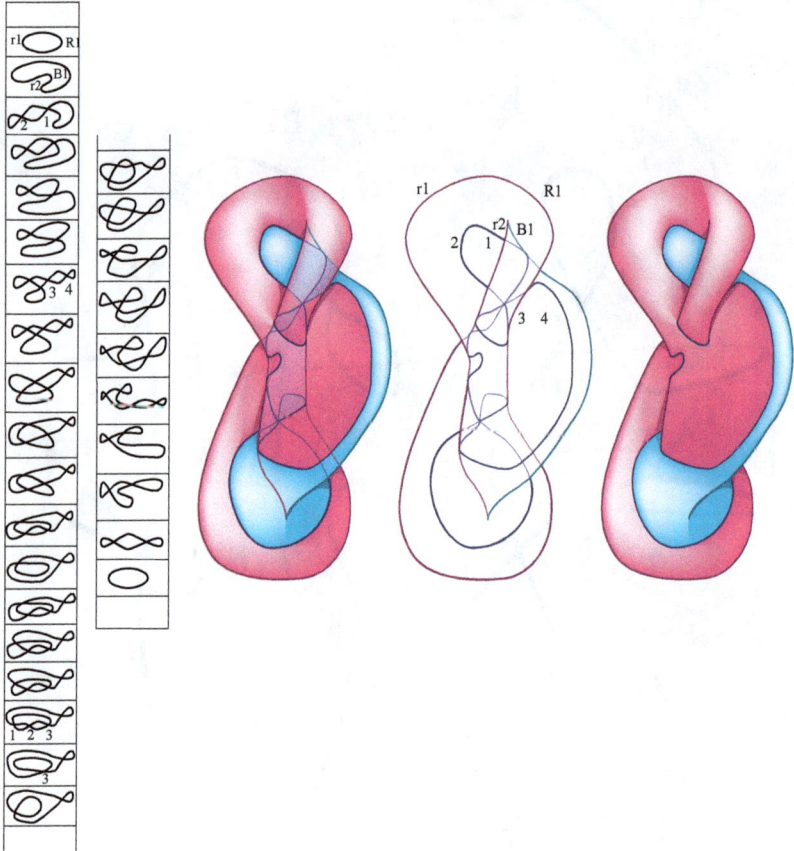

This is illustration red # 22. In the next stage, a type II zig-zag move occurs on the central tube, and a ψ, ψ-move cancels the two bounces between double point arc 1 and fold $r1$.

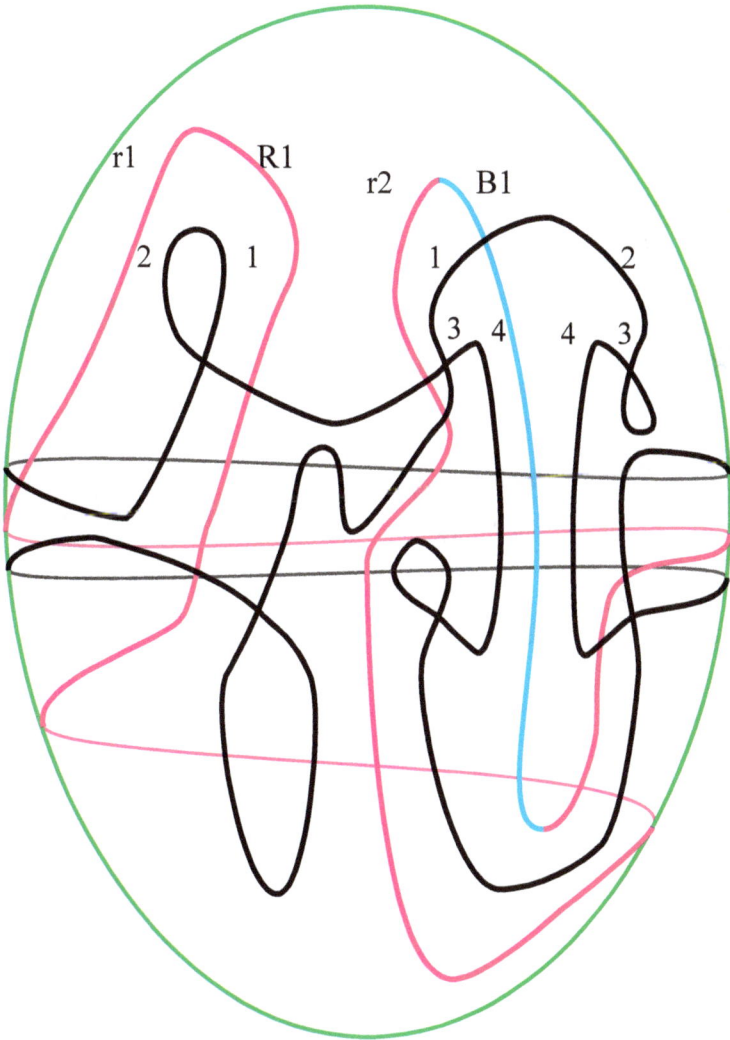

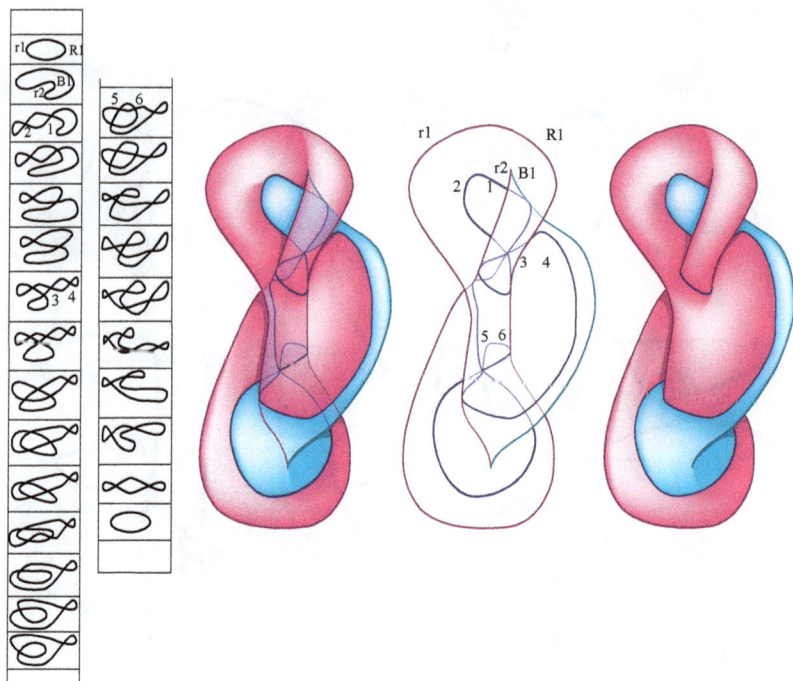

This is illustration red # 23. In the next stage, the type II birth of 1 and 3 (reading from the bottom) occurs below the death of 5 and 6 via a critical exchange. Both events lie between the triple points.

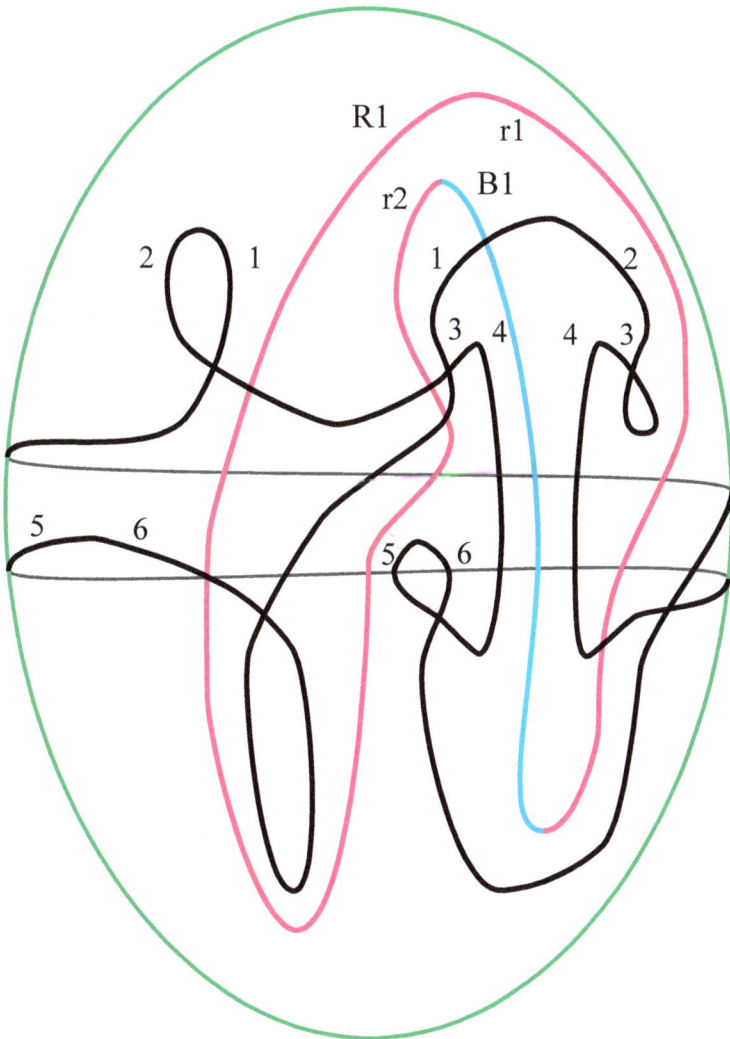

A re-ordering of the Gauss-Morse codes at the top and bottom has caused the red folds to swing over the top and bottom of the eggs.

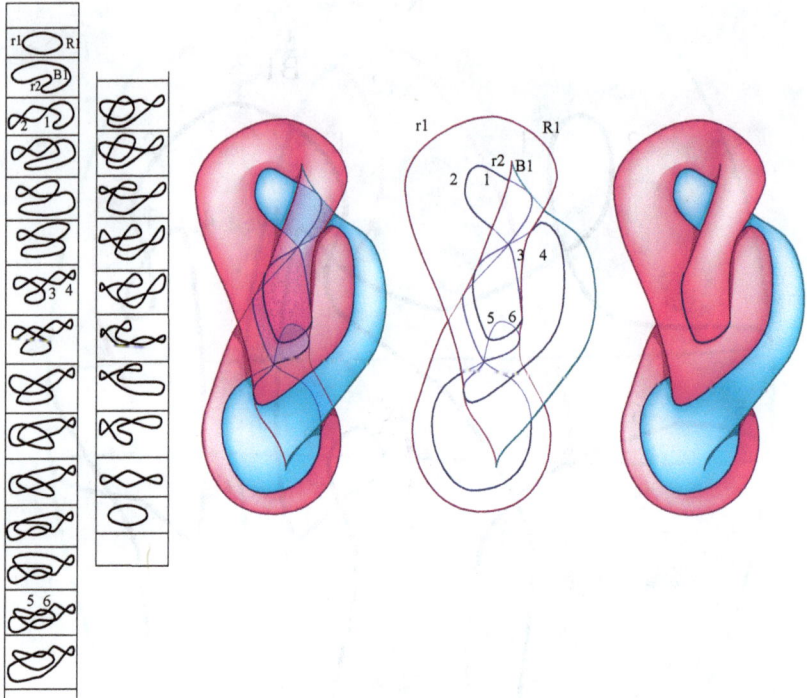

This is illustration red # 24. In the next stage, a type III-type III move occurs to create a pair of triple points and the bottom triple point bounces past the type II birth.

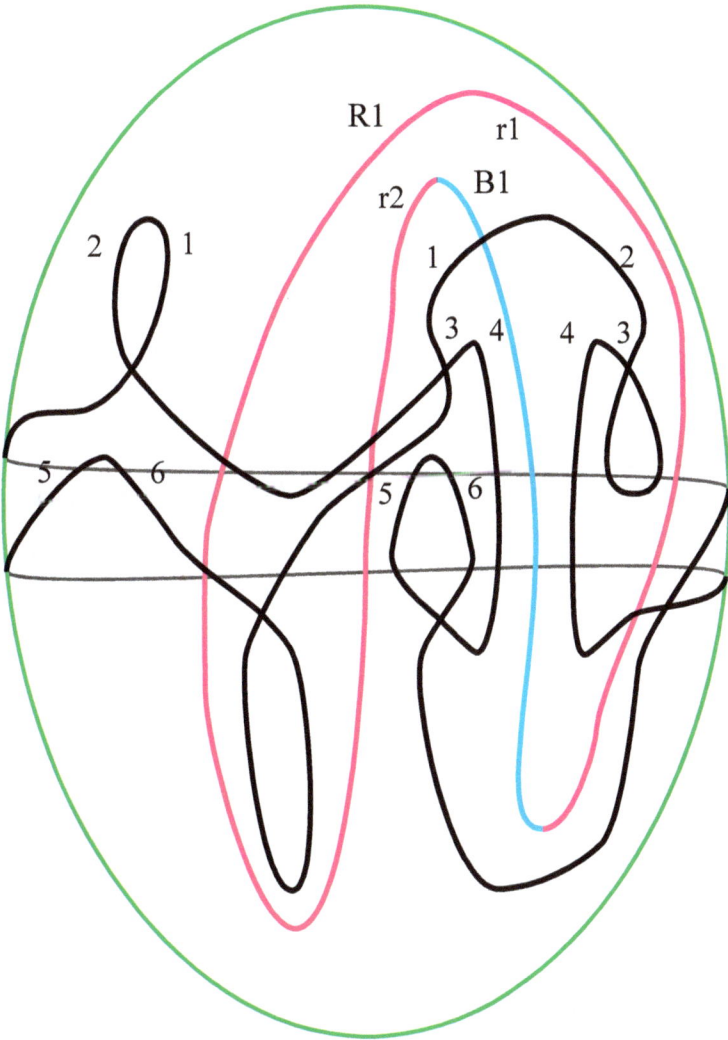

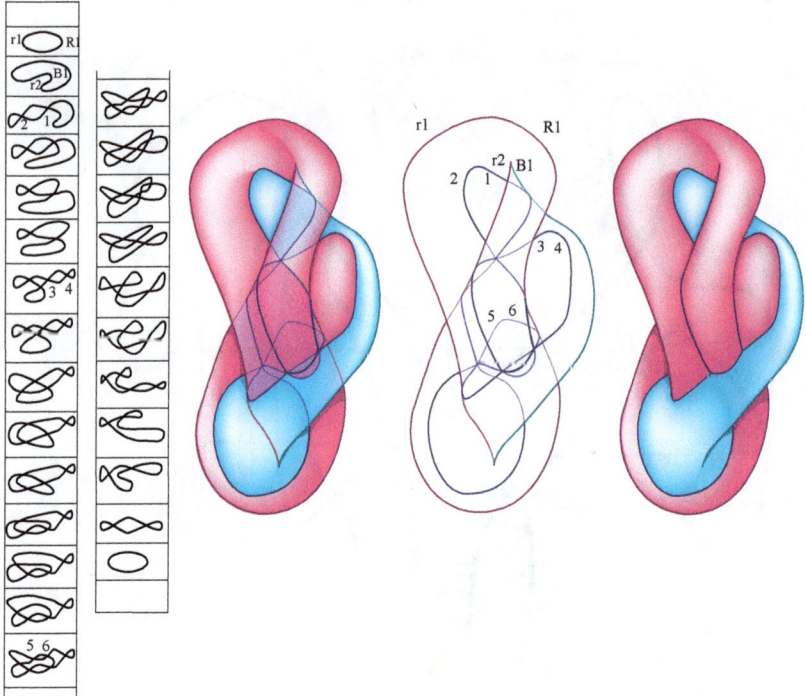

This is illustration red # 25. In the next stage, the triple point among double point arcs 2, 4, and 6 at the right bottom passes over the fold $R1$, and $R1$ stays to the far right before the death of double points 3 and 4.

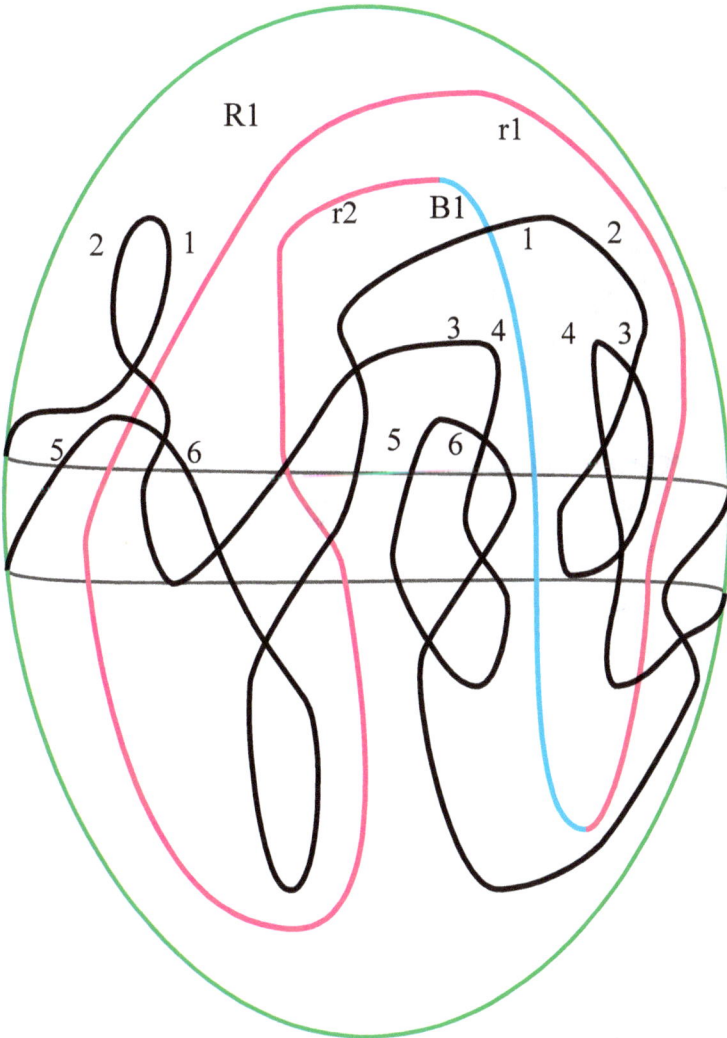

In the next decker set, the double point between arcs 1 and 5 moves from the right side of the egg to the left side. This double point corresponds to the 1, 5, 6 triple point that is the bottom-most.

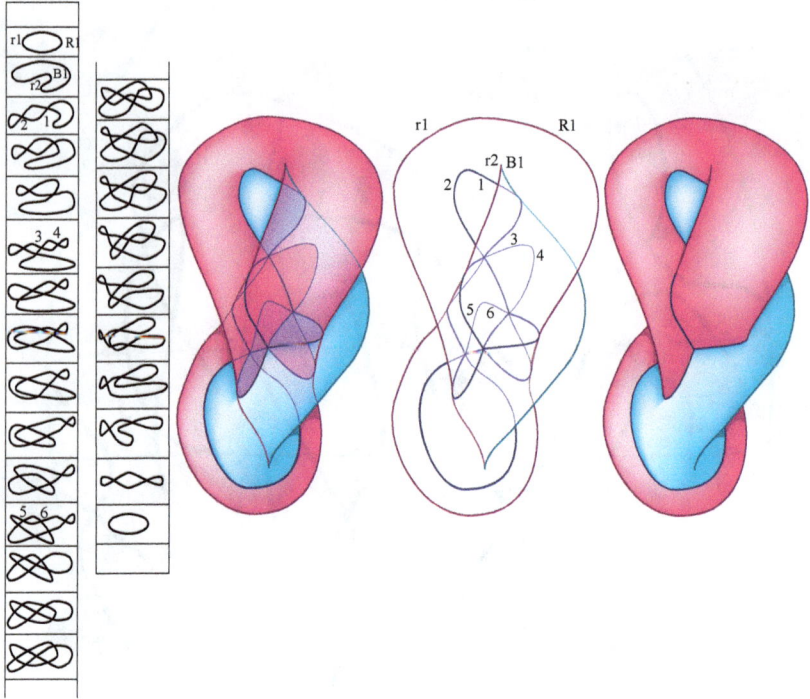

This is illustration red # 26. In the next stage, there is a triple point bounce between the death of double point arcs 5 and 6. The triple point 246 becomes 245.

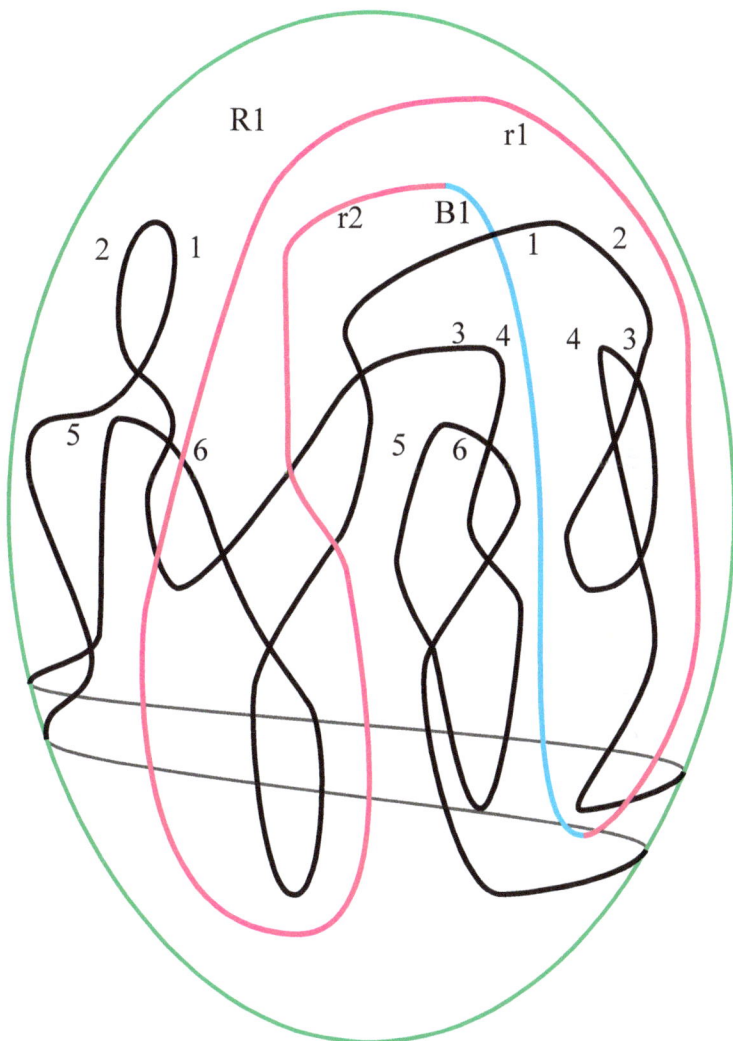

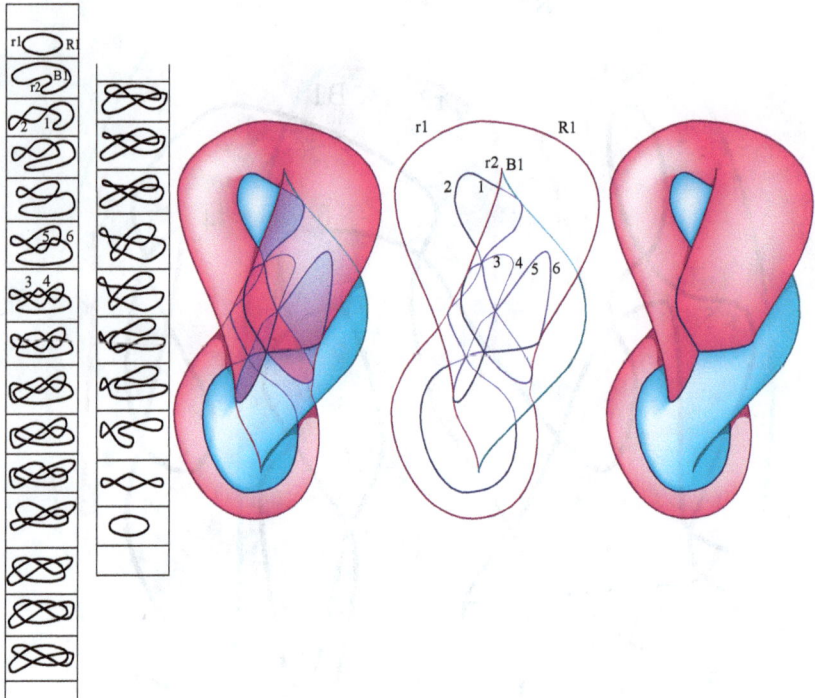

This is illustration red # 27. In the next stage, triple point 123 passes to the front of the blue fold $b1$.

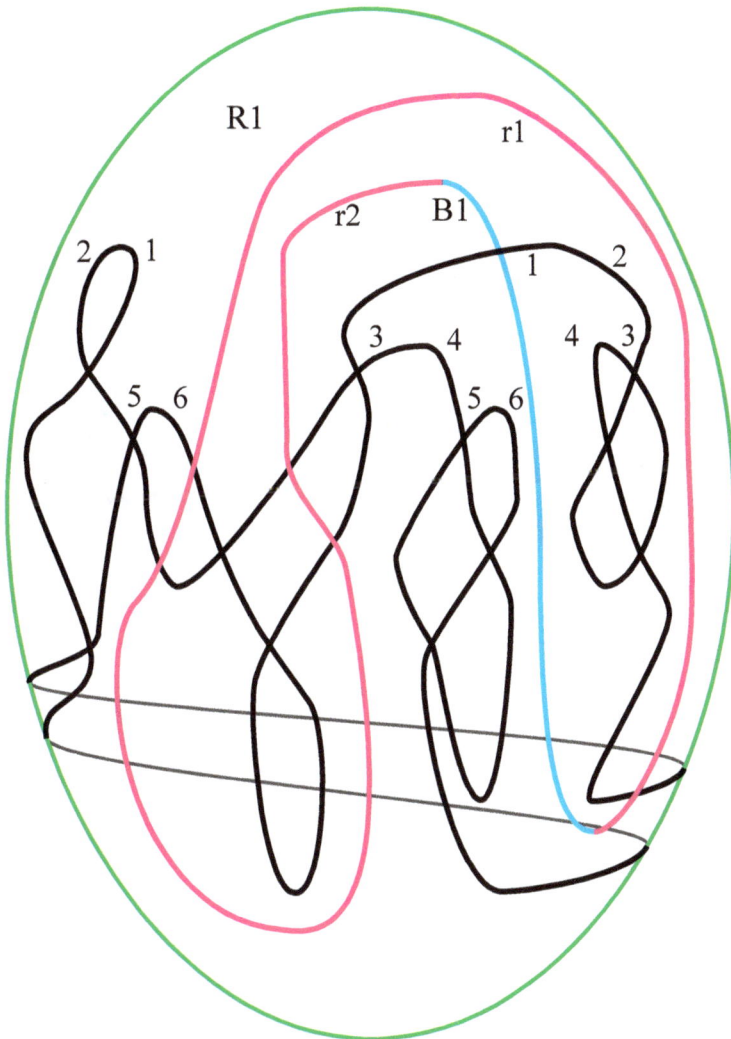

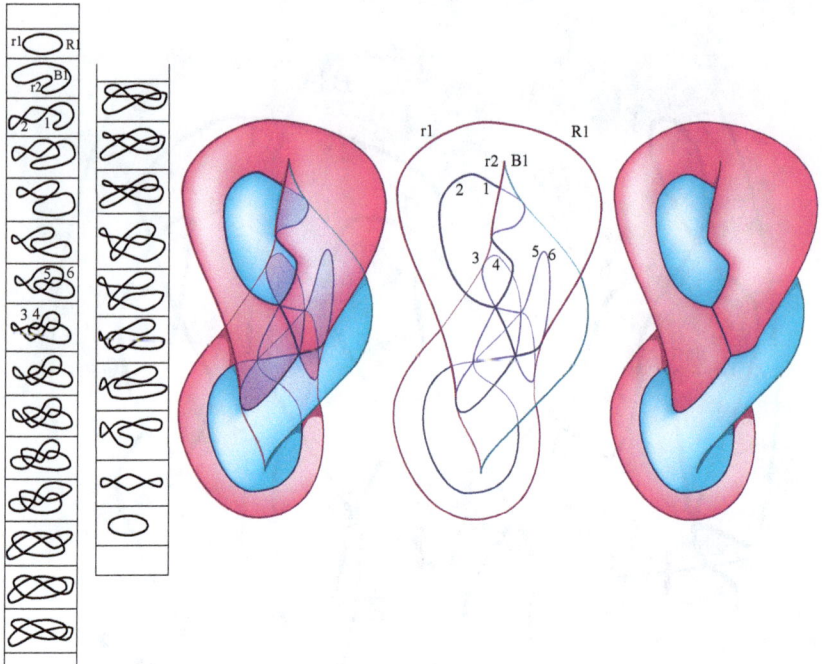

This is illustration red # 28. In the next stage, the quadruple point move occurs among the 4 triple points, and the tetrahedron in the center of the figure appears to turn inside-out. At that time, the sphere has turned blue.

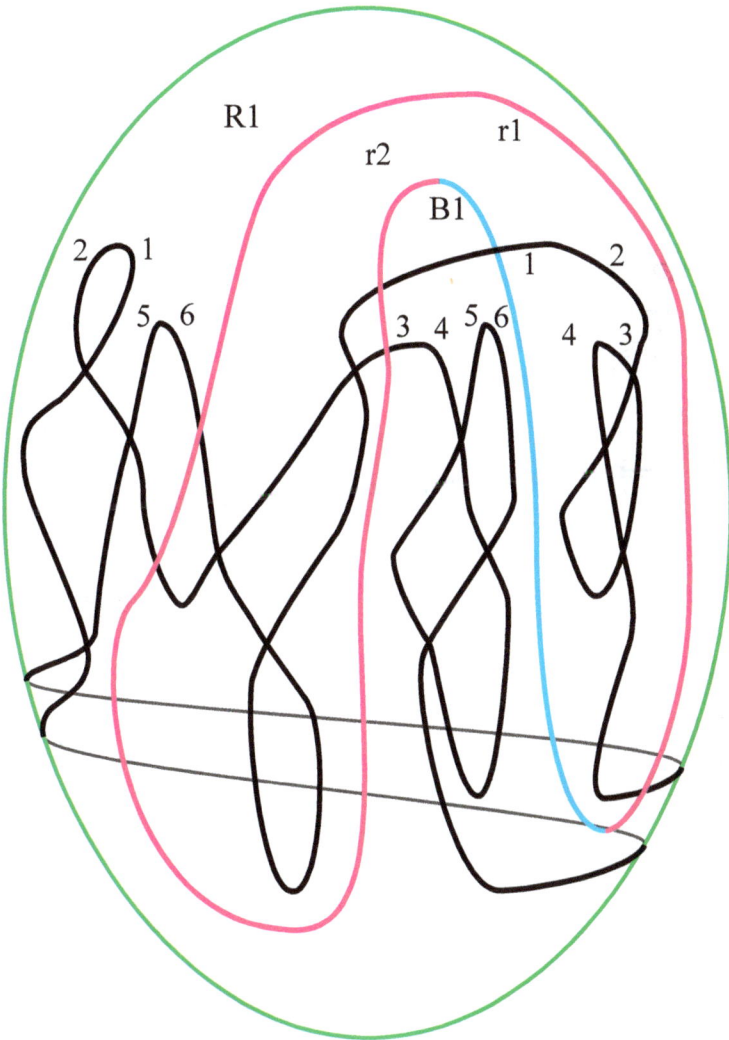

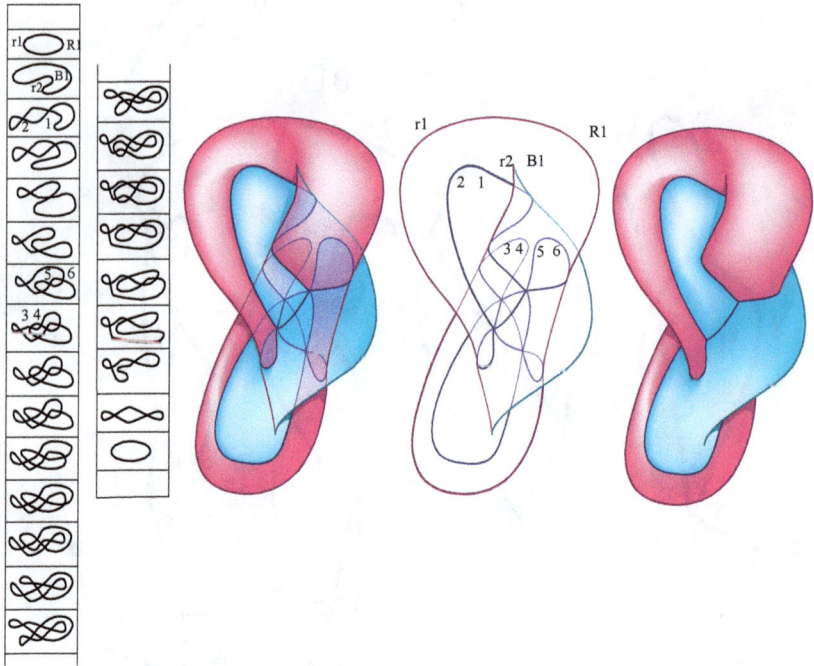

This is illustration blue # 53. In the next stage, double point arc 6 bounces over the fold $B1$ via a ψ, ψ-move.

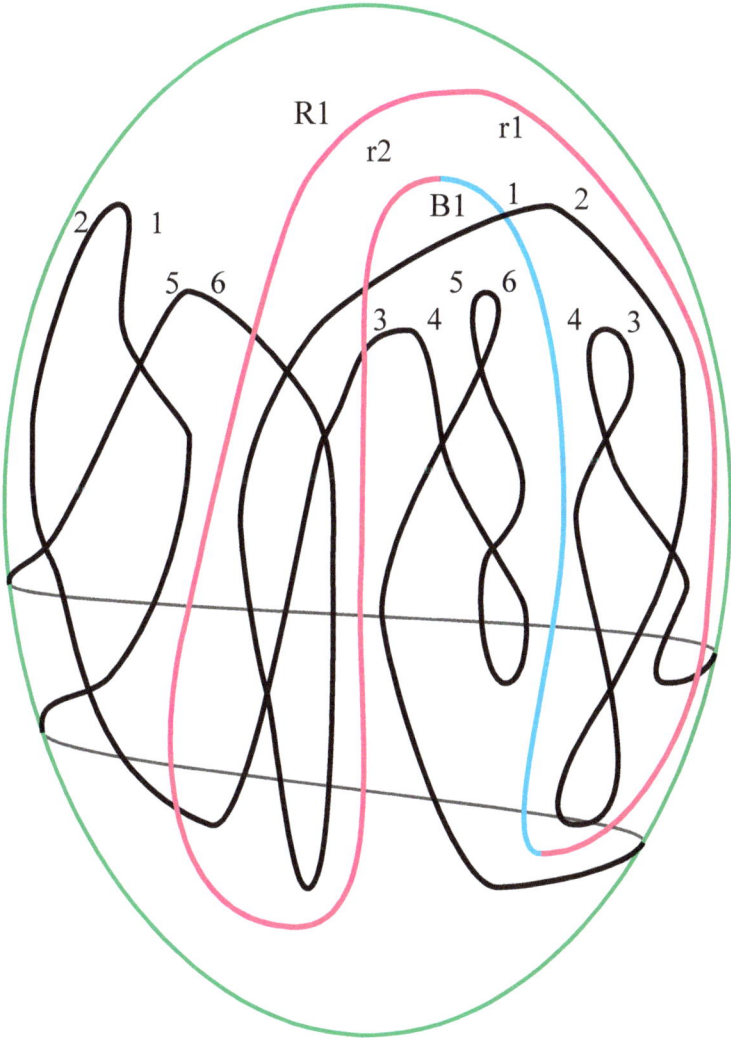

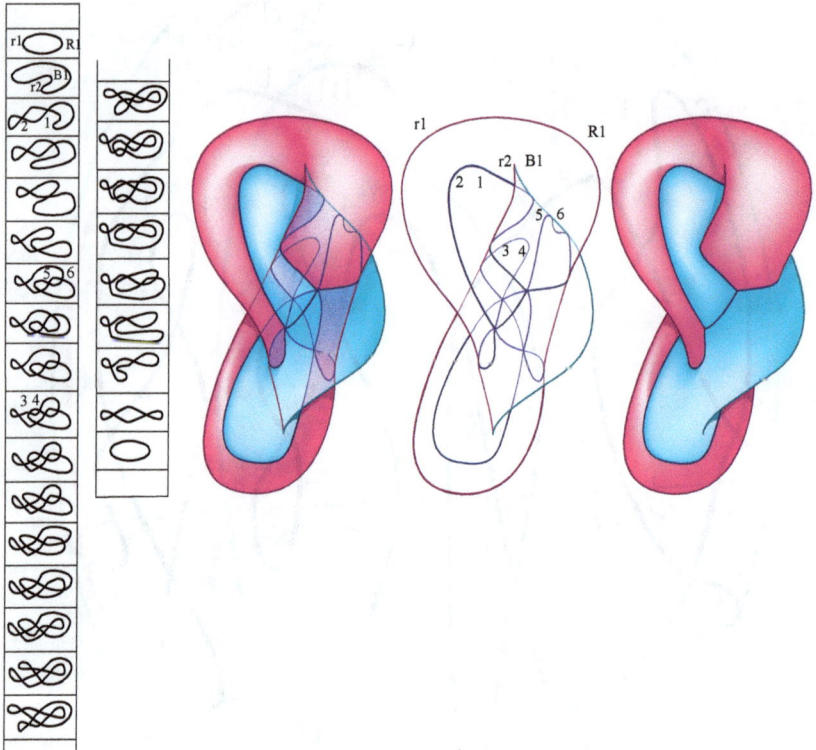

This is illustration blue # 52. In the next stage, the combination of a pair of type II zig-zag moves and a type II saddle occurs to cancel the min towards the top of the figure, and the Max towards the bottom.

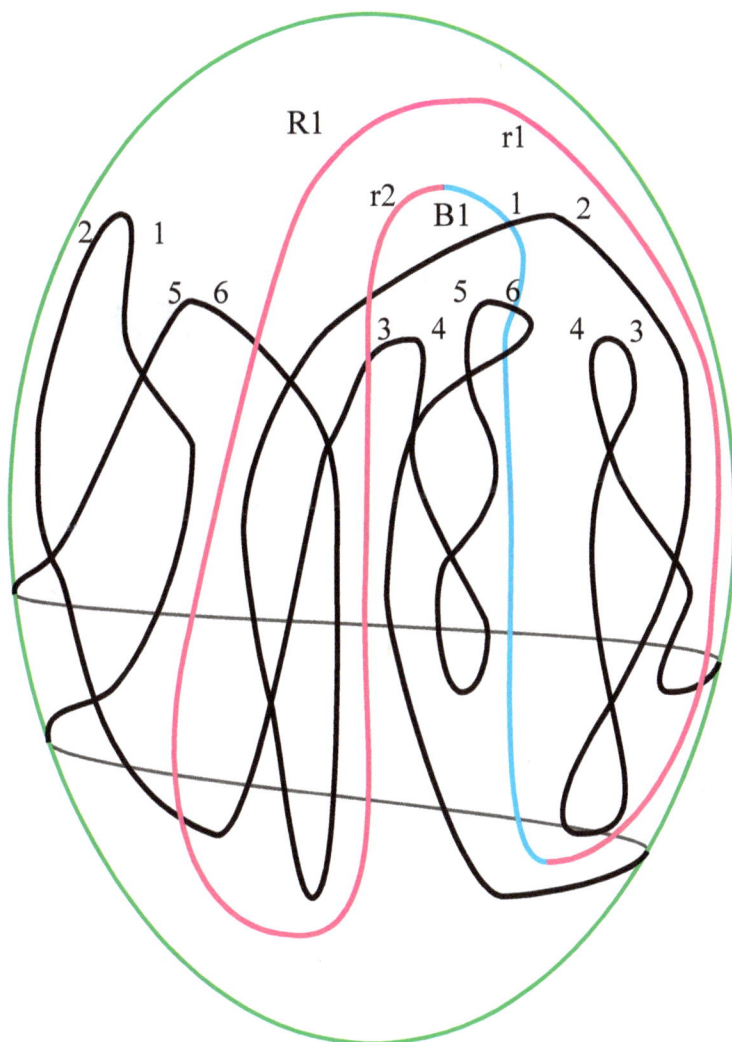

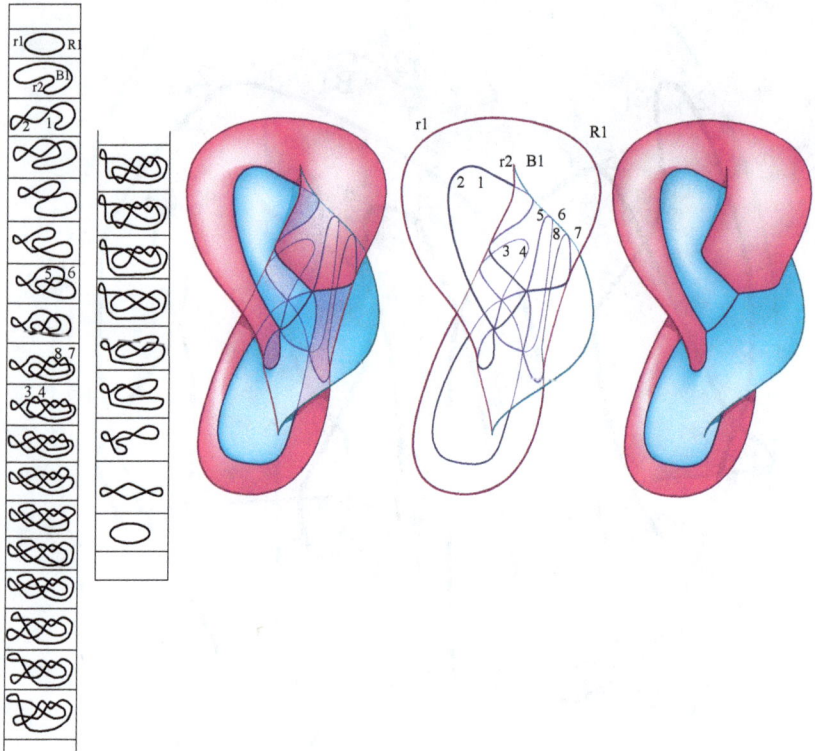

This is illustration blue # 51. In the next stage, a horizontal type II move occurs among double point arc 7, the fold $B1$, and the fold $R1$. The resulting two veiled bounces between 7 and $R1$ are canceled via a ψ, ψ-move.

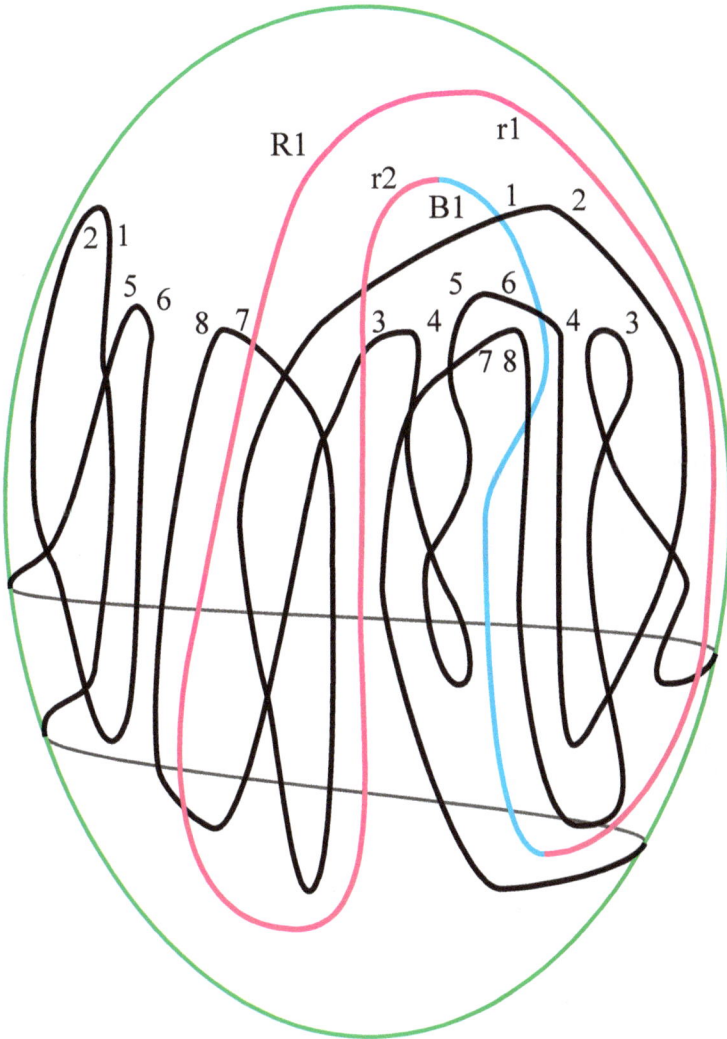

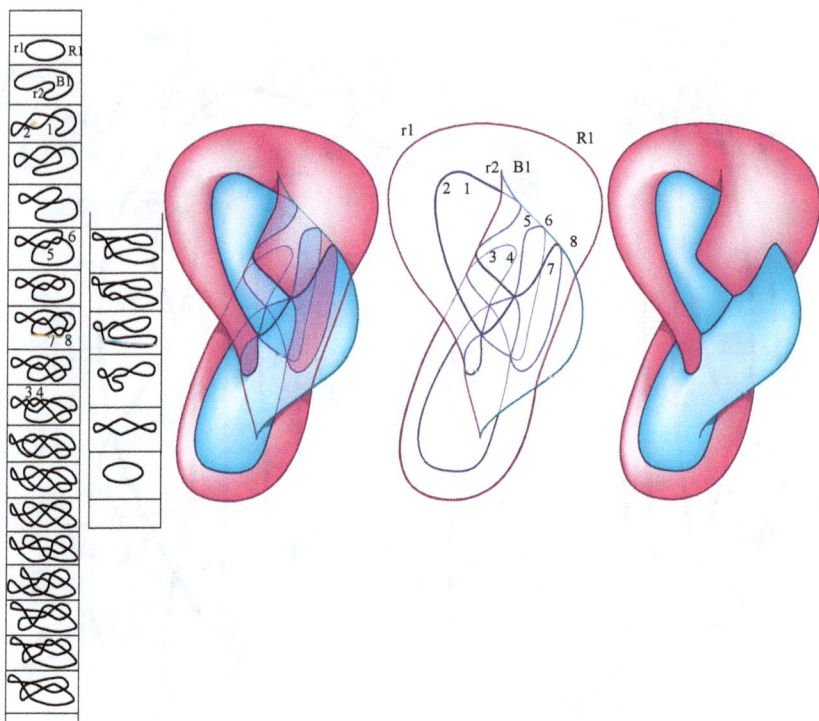

This is illustration blue # 50. In the next stage, the double point arc 1 moves to the other side of the top red cusp, and the pair of bounces between 1 and $r1$ are canceled by a ψ, ψ-move.

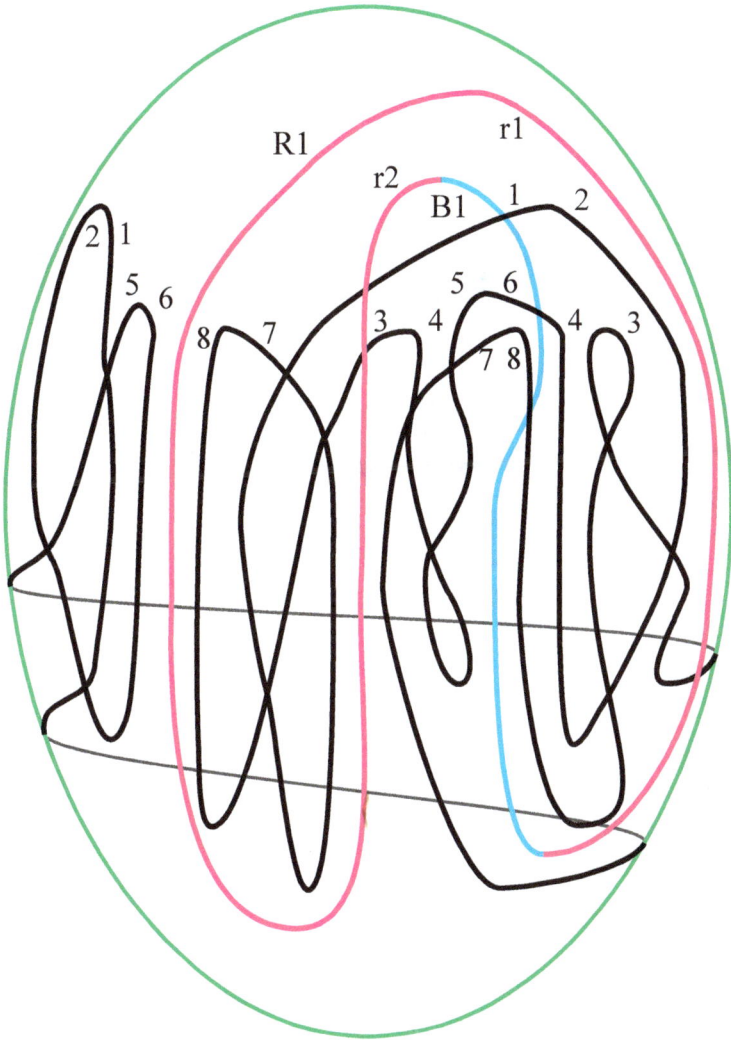

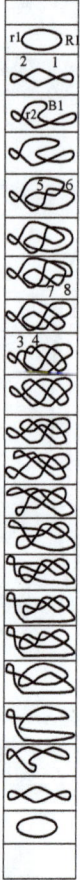

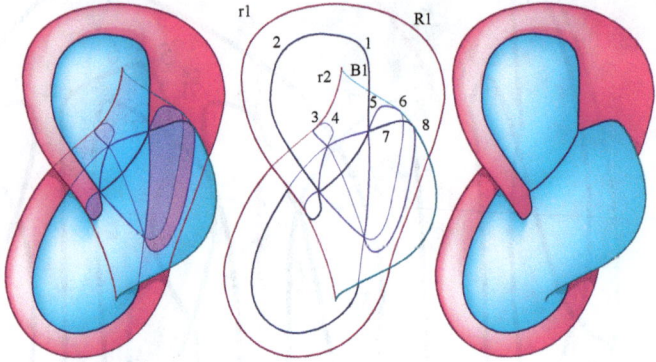

This is illustration blue # 49. In the next stage, the triple point 157 bounces to the right of the death of 7 and 8 to become 158.

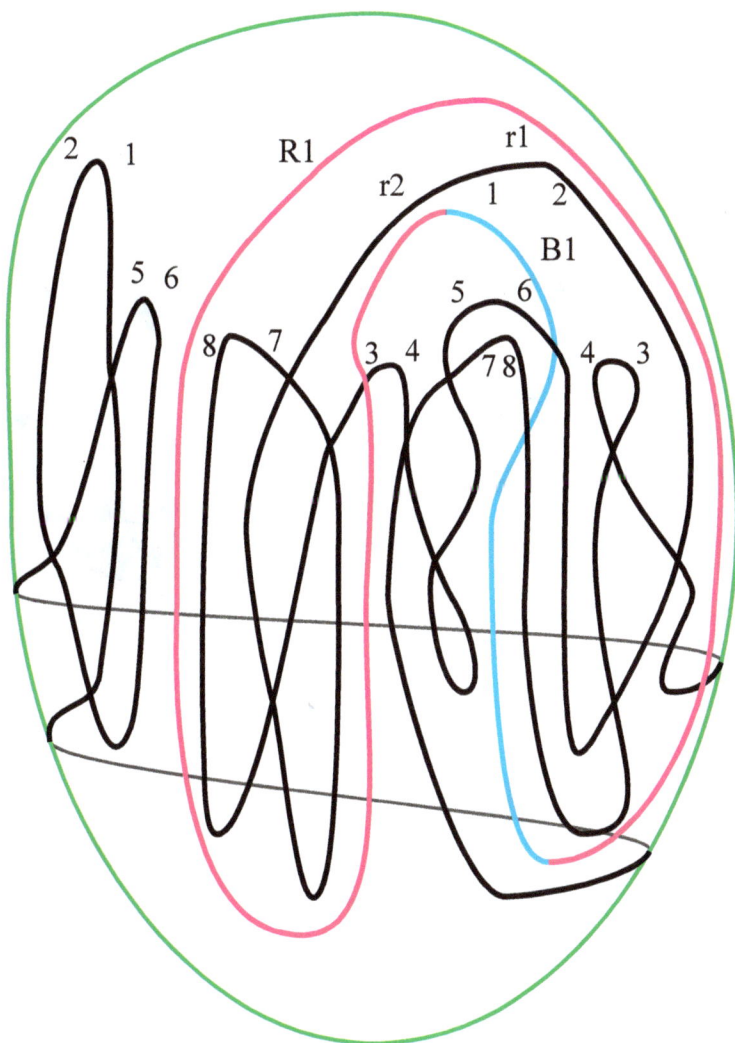

This is illustration blue # 48. In the next stage, the death of 7 against 8 moves above the death 5 against 6.

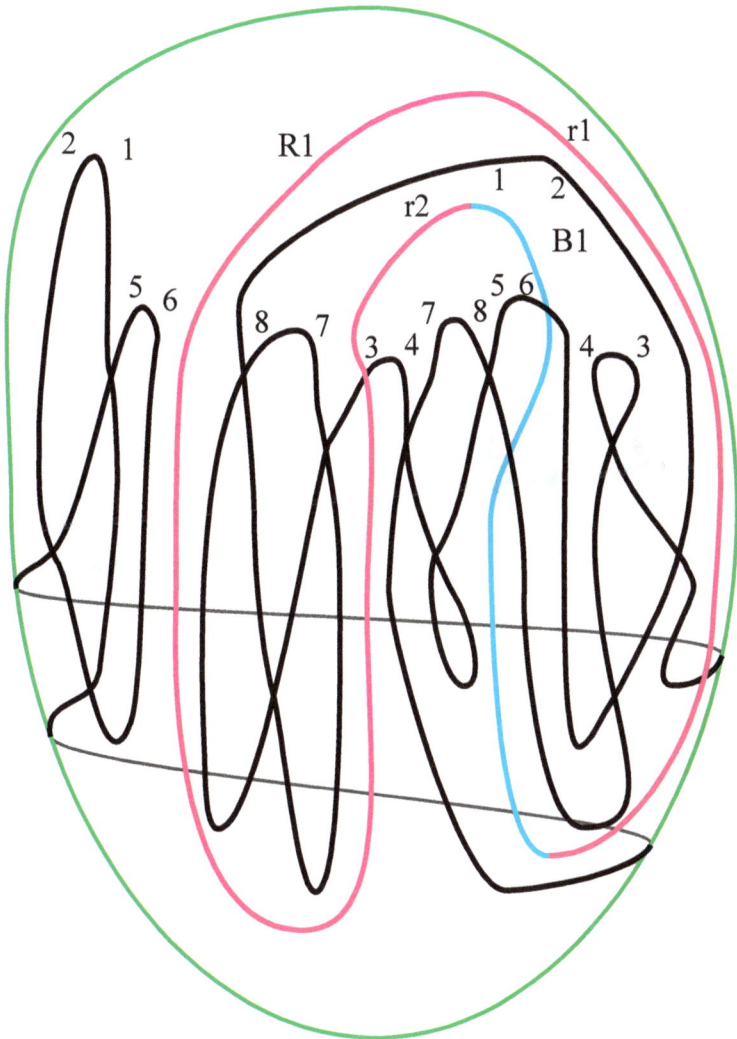

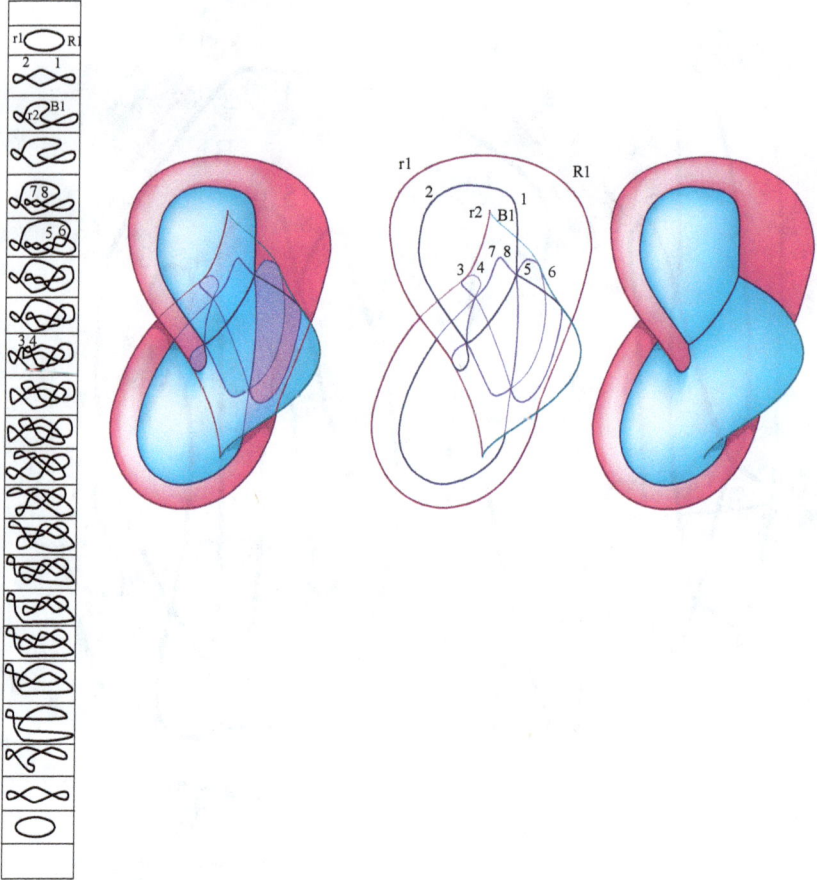

This is illustration blue # 47. In the next stage, the triple point 158 bounces to the right of the death of 5 and 6 to become 168.

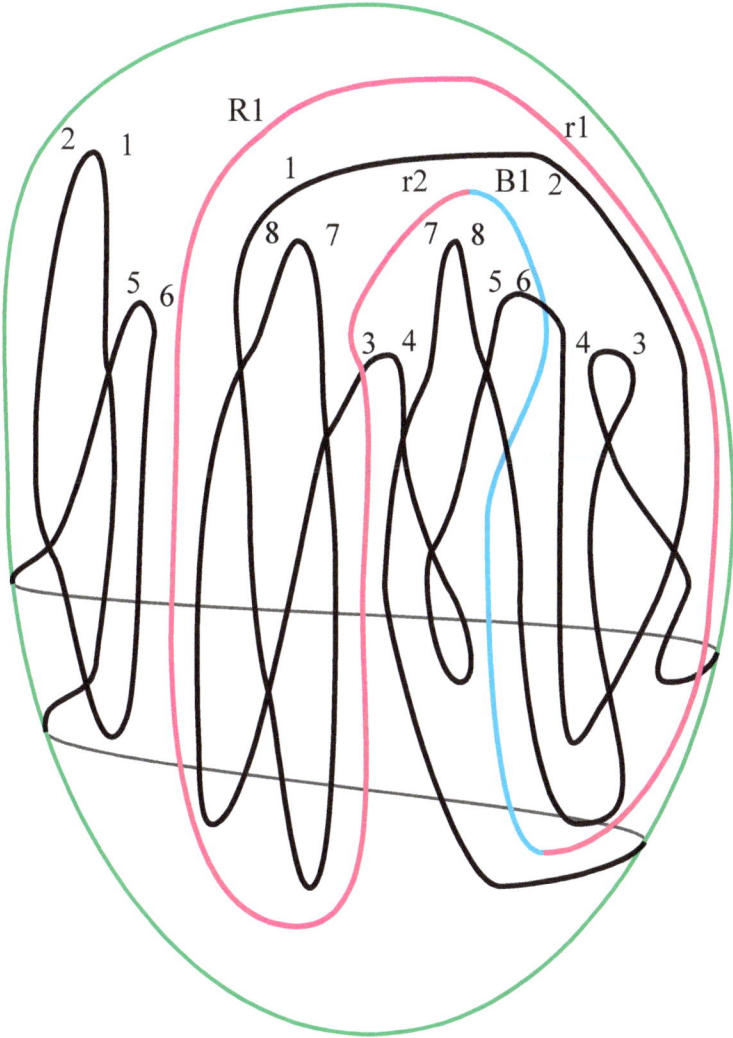

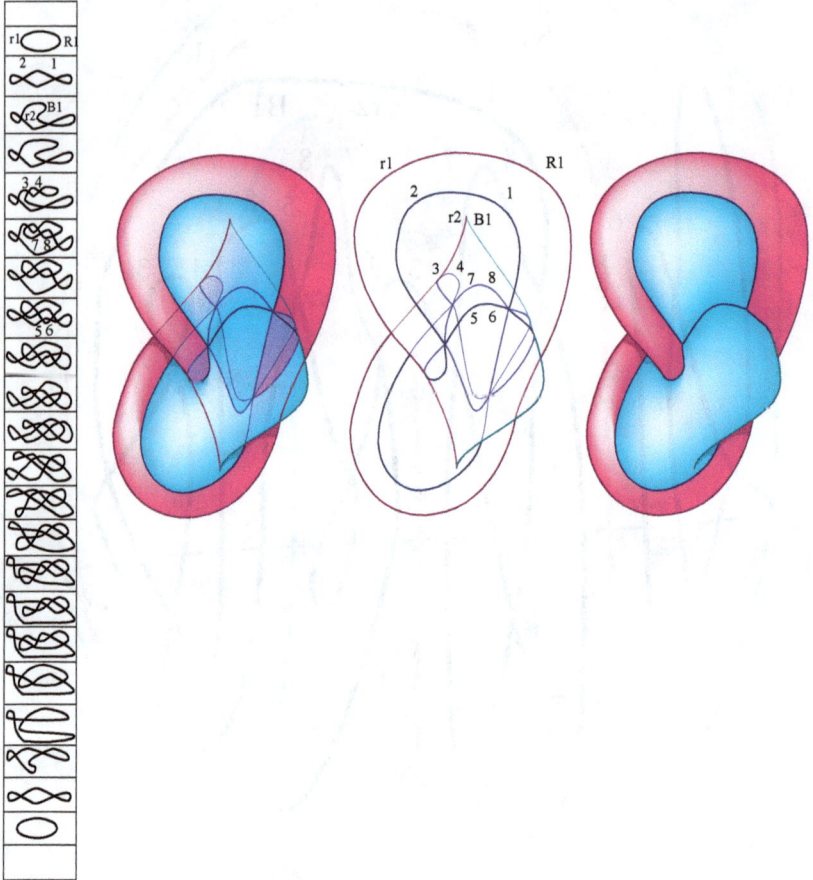

This is illustration blue # 46. In the next stage, the pair of triple points on the right of the illustration of the form 168 cancel via a type III-type III move.

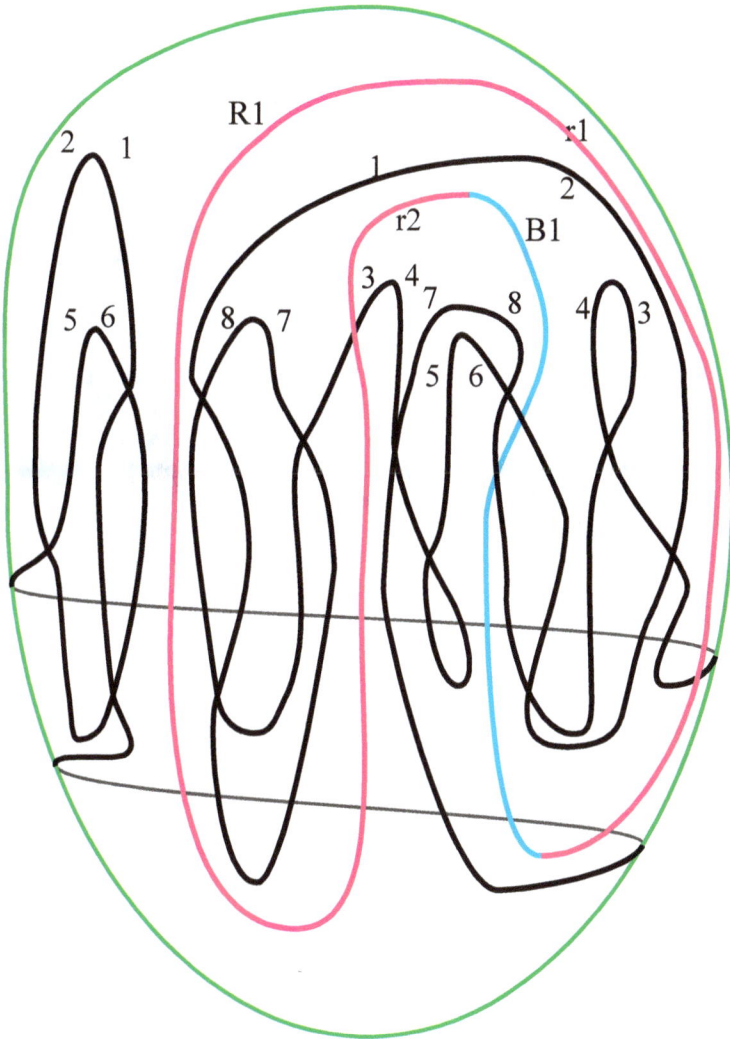

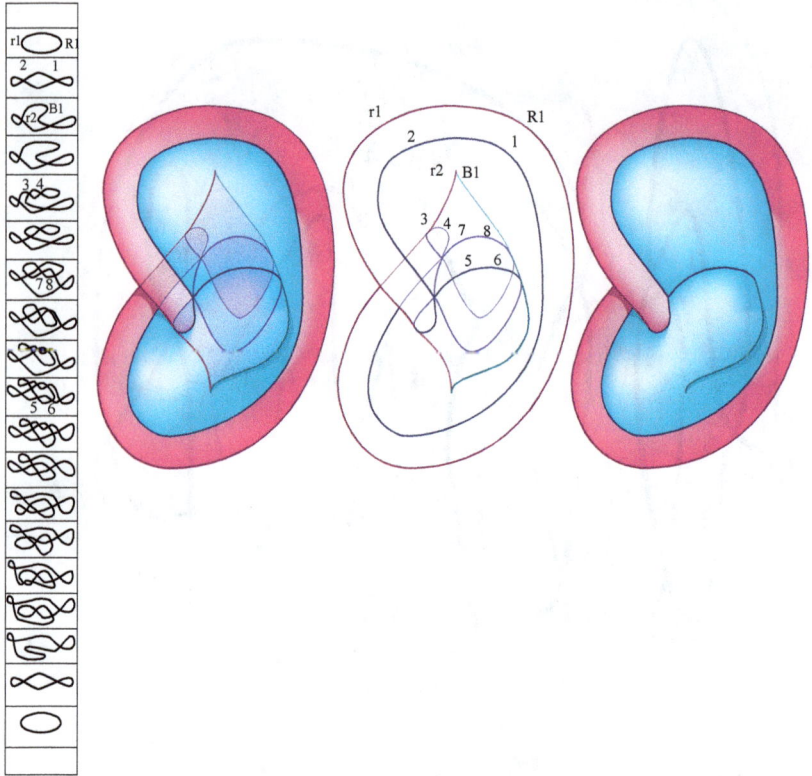

This is illustration blue # 45. In the next stage, the death of 5 against 6 moves below the birth of 3 and 8, and a lips occurs to induce a visible left blue fold after the birth of 1 and 7.

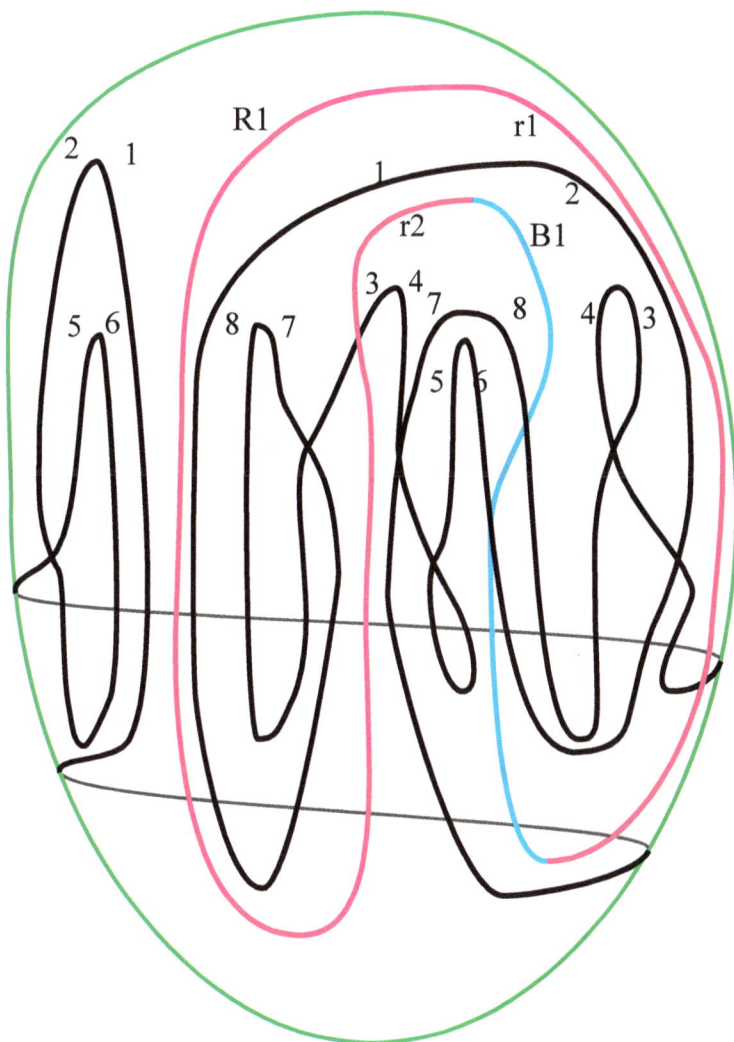

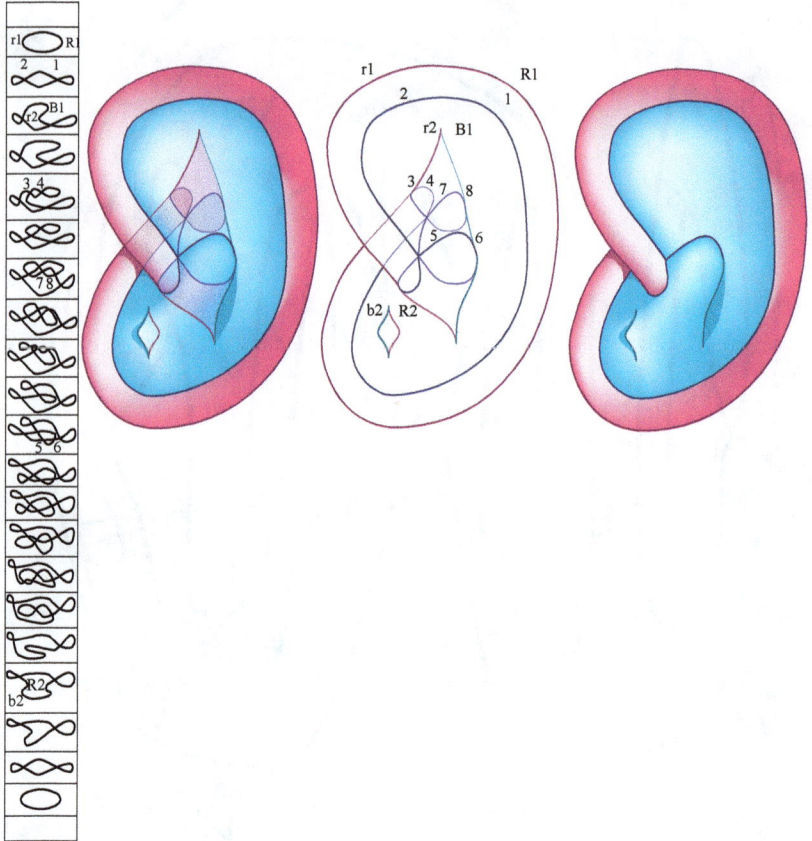

This is illustration blue # 44. In the next stage, the top cusp between $b2$ and $R2$ moves up the surface towards the triple point 347. The blue fold $b2$ moves to the left of the double curve 7 and the red fold r2.

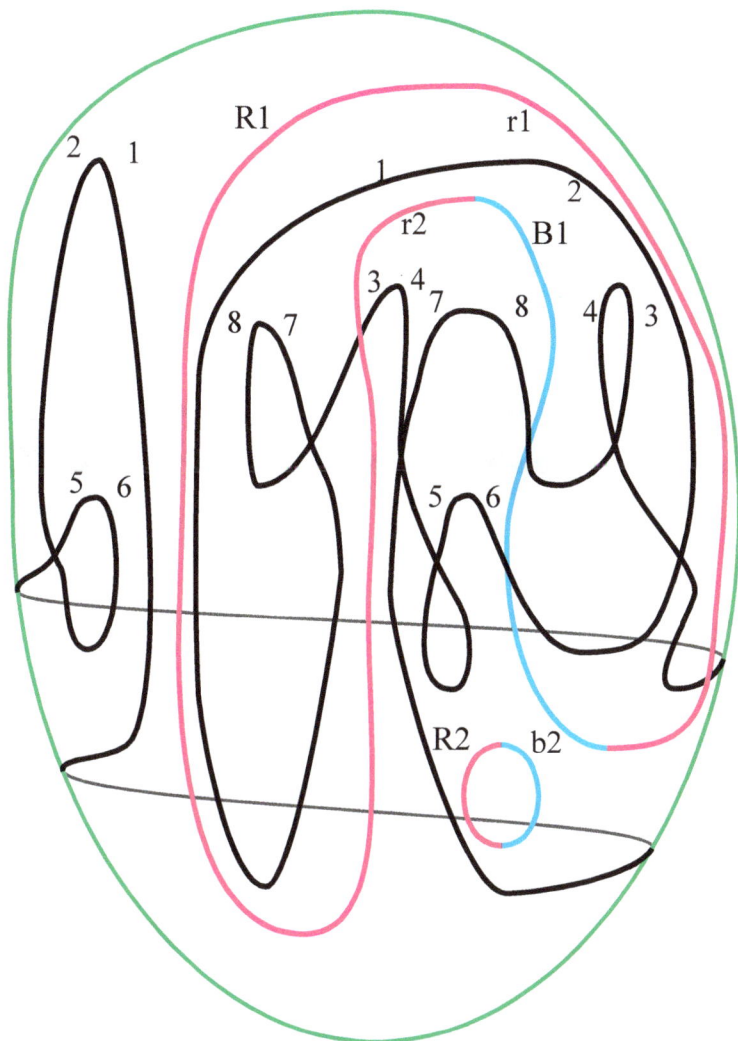

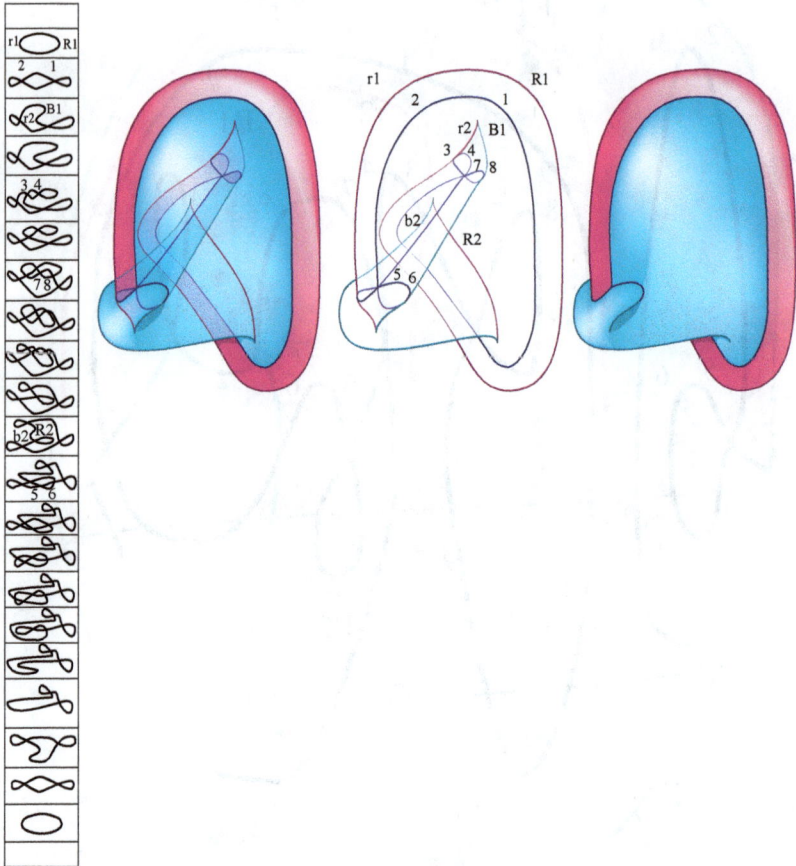

This is illustration blue # 43. In the next stage, a pair of canceling critical points (saddle and minimum) is created on the blue arc $b2$ and the minimum is pushed to the bottom of the figure. The saddle is visible and represents the birth of fold arcs $B2$ on the left and $b3$ on the right.

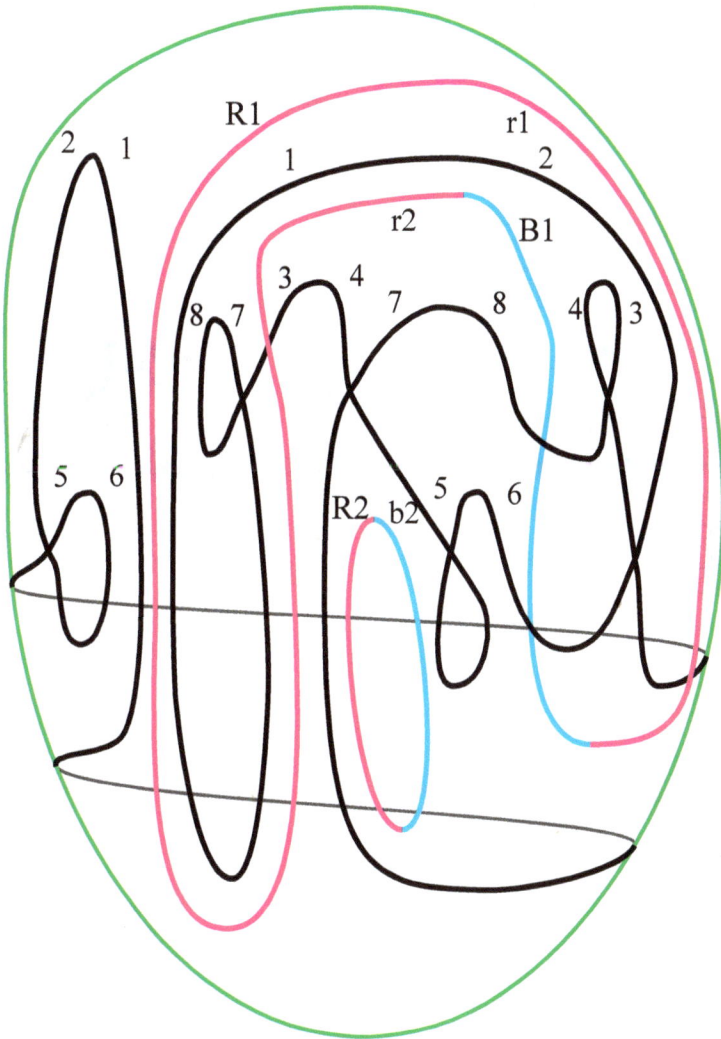

Watch the decker set in the next illustration carefully. Items on the left move around the back to reappear on the right. A saddle point/minimum pair of critical points is created towards the bottom. The items on the right pendent form the 245 triple point on the blue foot that protrudes on the left on the corresponding chart.

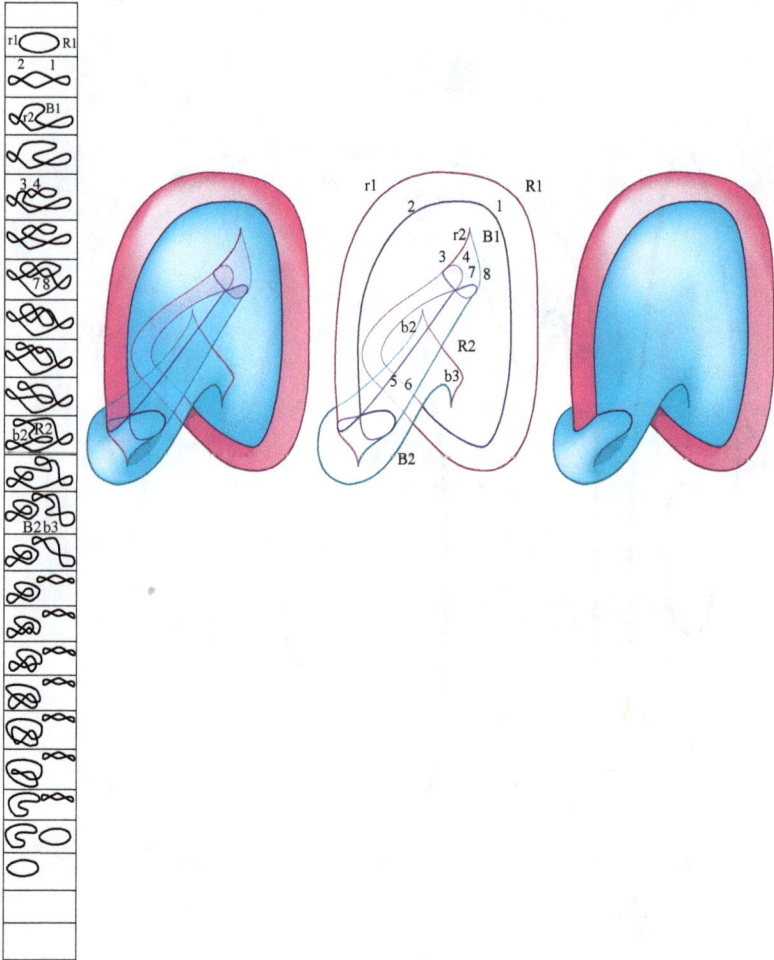

This is illustration blue # 42. In the next stage, a horizontal cusp change occurs along the cusp $R2/b3$, and this event is pushed towards the top of the illustration. Blue fold $B2$ veils portions on the left, and other critical levels are interchanged.

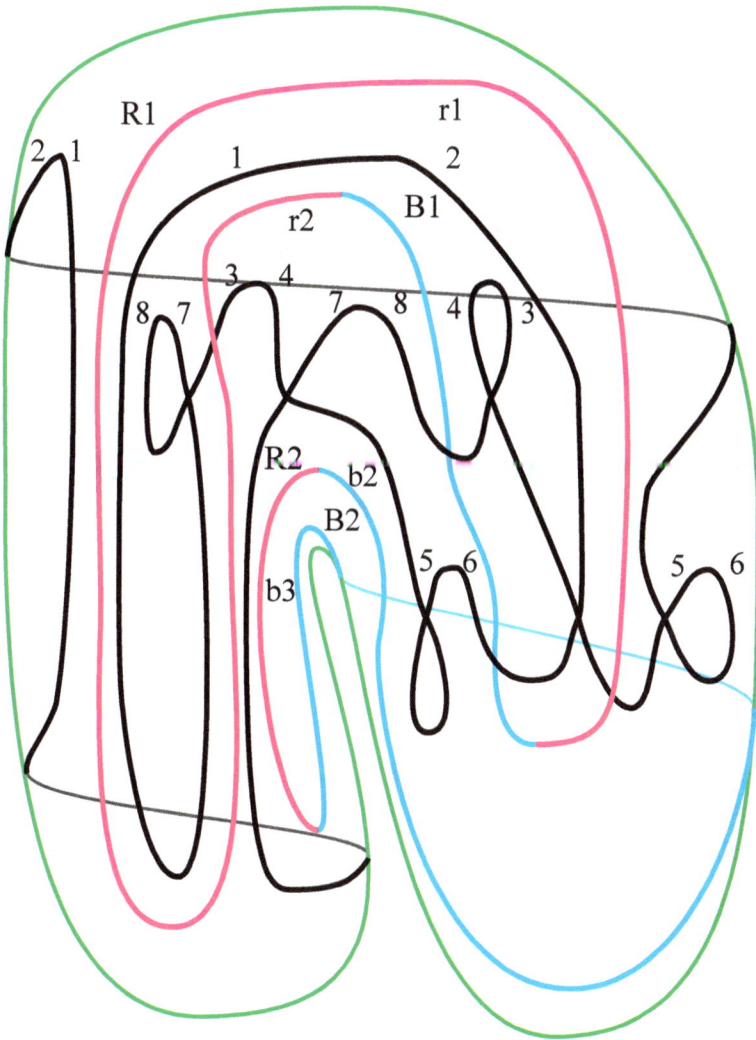

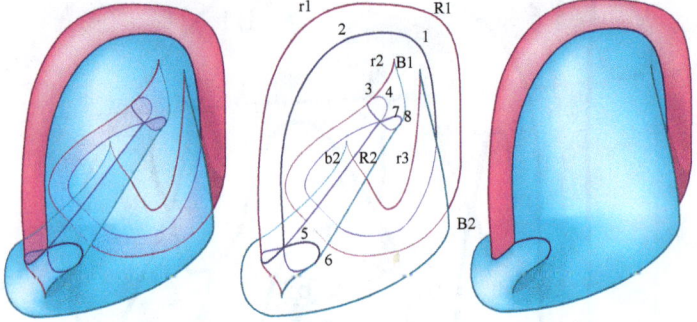

This is illustration blue # 41. In the next stage, a ψ, ψ-move occurs between the double point curve 1 and the fold $r3$.

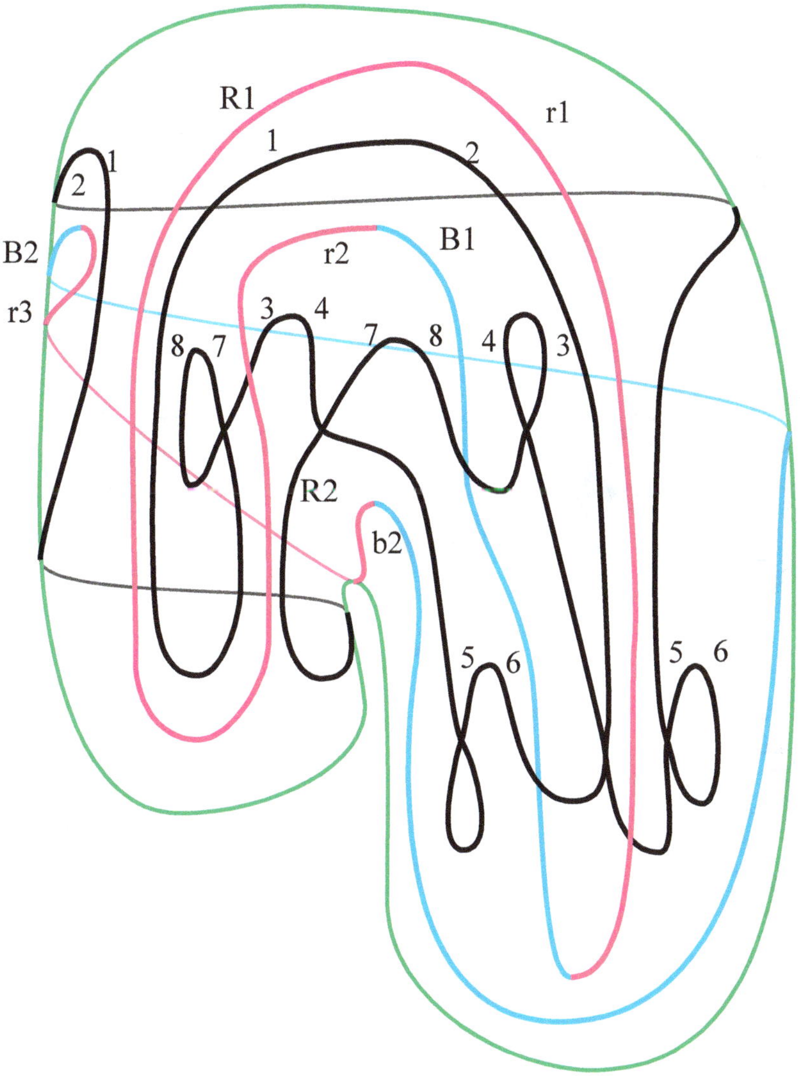
R1
r1
1
2
2
1
B2
r2
B1
r3
3
4
8
7
7
8
4
3
R2
b2
5
6
5
6

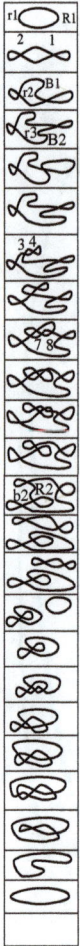

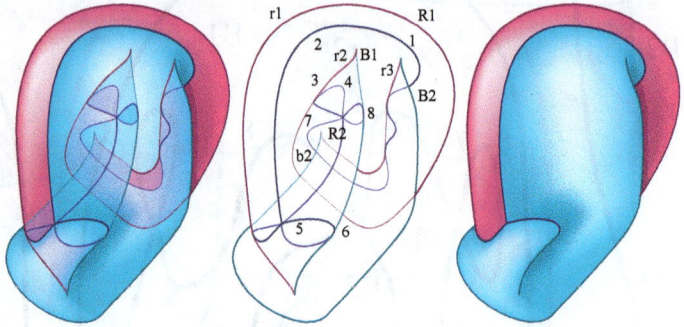

This is illustration blue # 40. In the next stage, the type II move birth of 17 passes through the saddle point that is the birth of $R2, r3$.

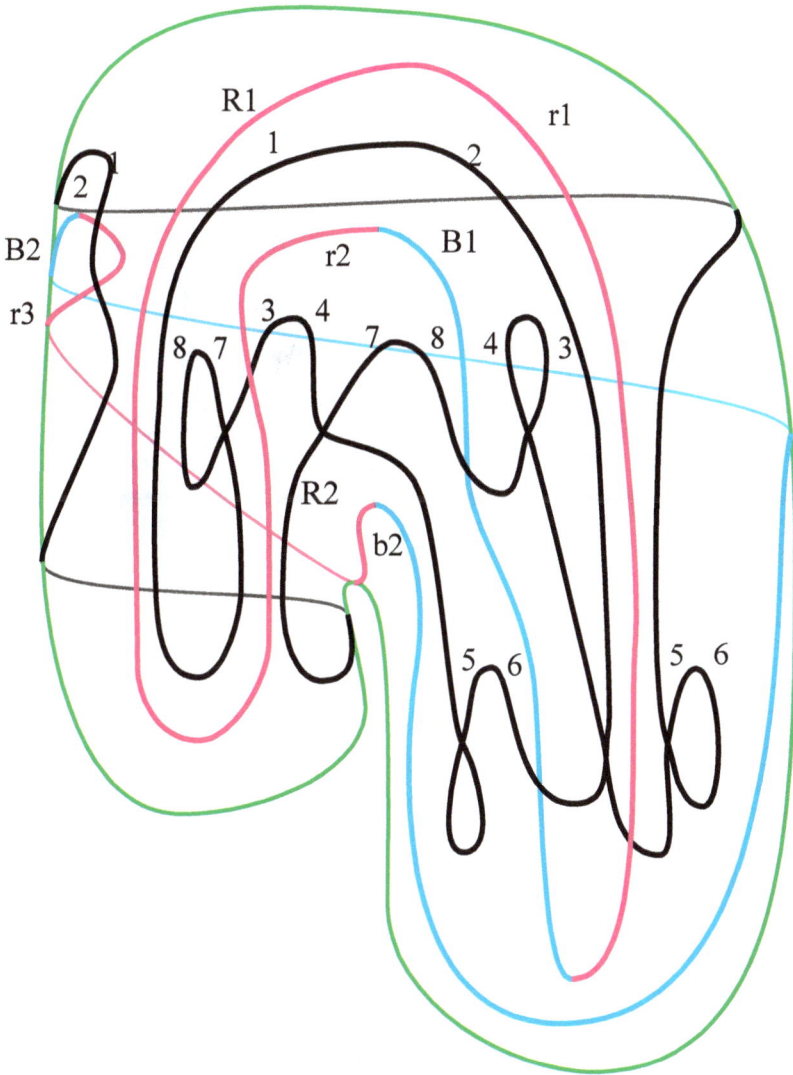

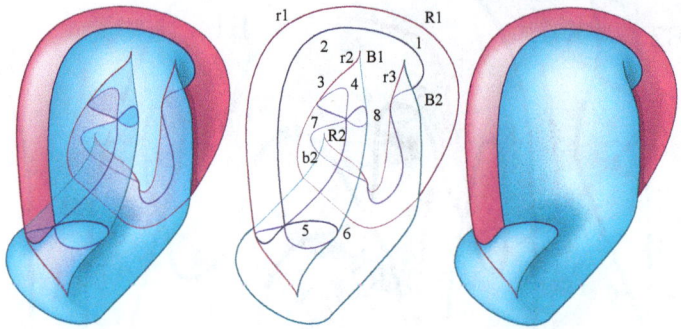

This is illustration blue # 39. In the next stage, the ψ-move between the double point arc 7 and the red fold $R2$ moves through the cusp.

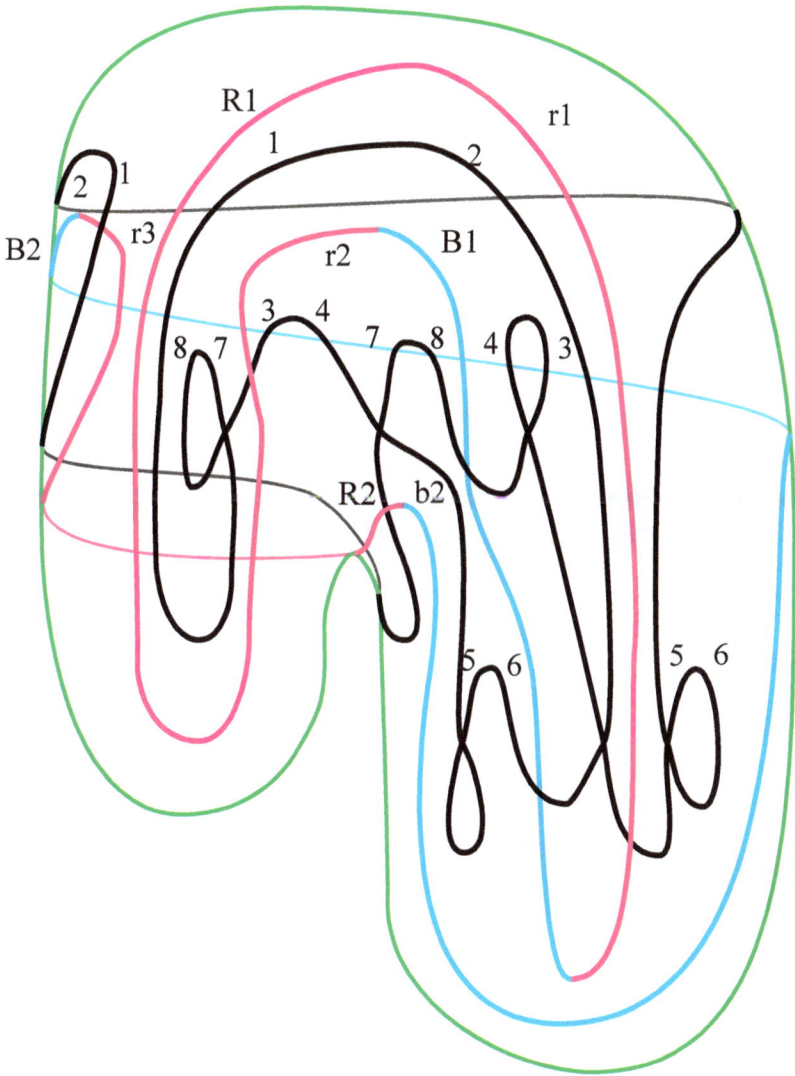

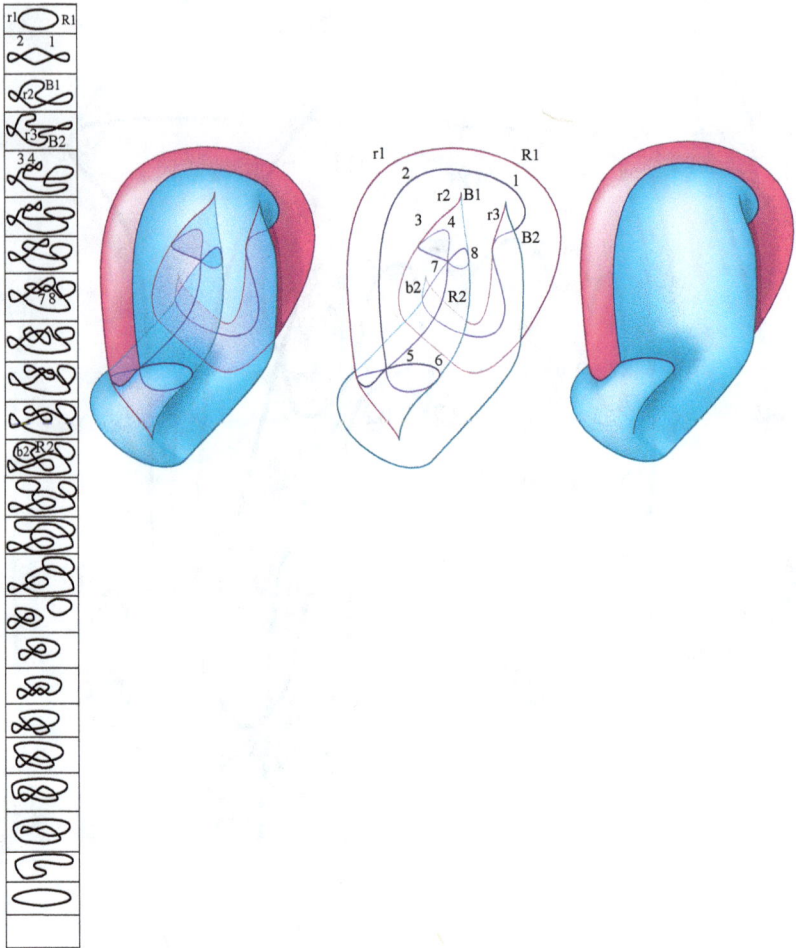

This is illustration blue # 38. In the next stage, a canceling pair of critical points is added along red fold $R2$, the maximum moves above the cusp $b2/R2$, and a horizontal cusp occurs between this cusp and the newly introduced folds at the saddle points. The new critical points are labeled in the movie and the decker set but not in the chart since there is little space to indicate them in this figure.

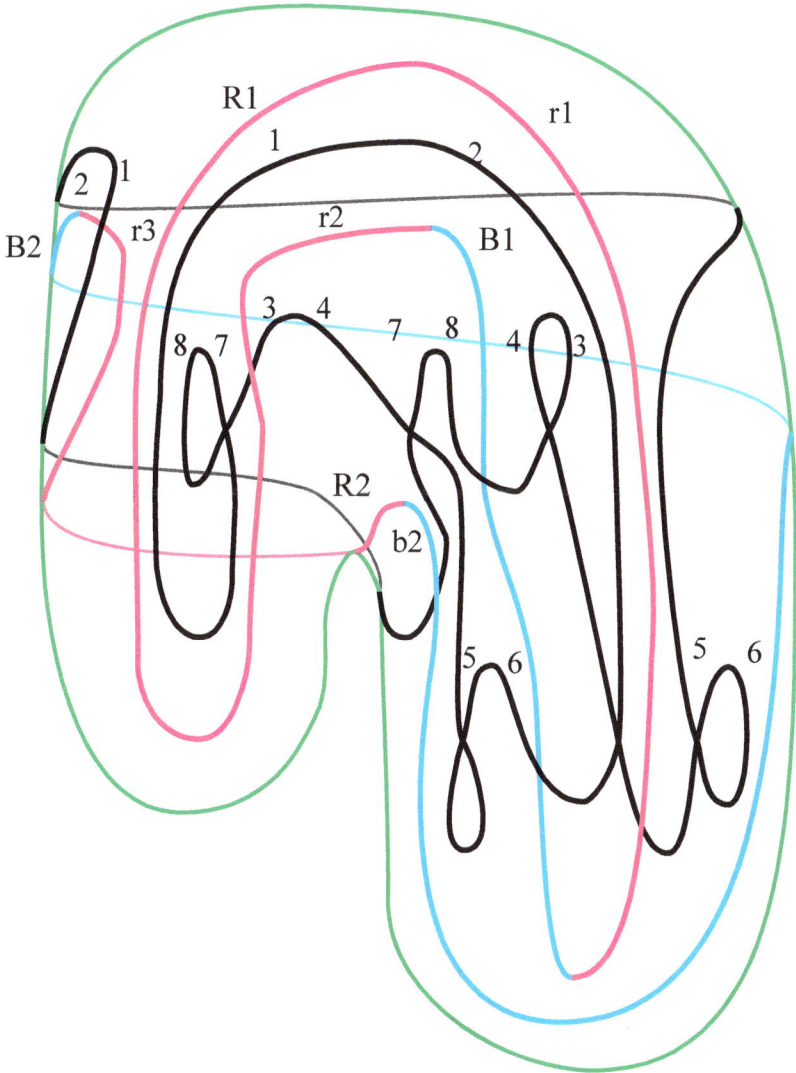

In the next decker set, arcs 1, 2, *B*2, *r*3, and *R*1 are passed to the right of the figure, and a pair of canceling critical points are added to the ambient sphere.

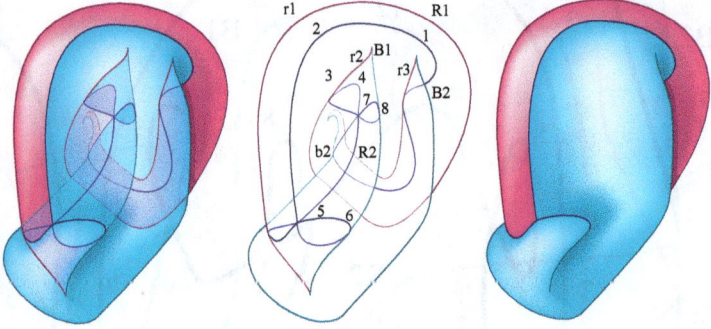

This is illustration blue # 37. In the next stage, a swallow-tail is added on the red arc *r2*. There is no need to indicate the name of the blue arc that is introduced.

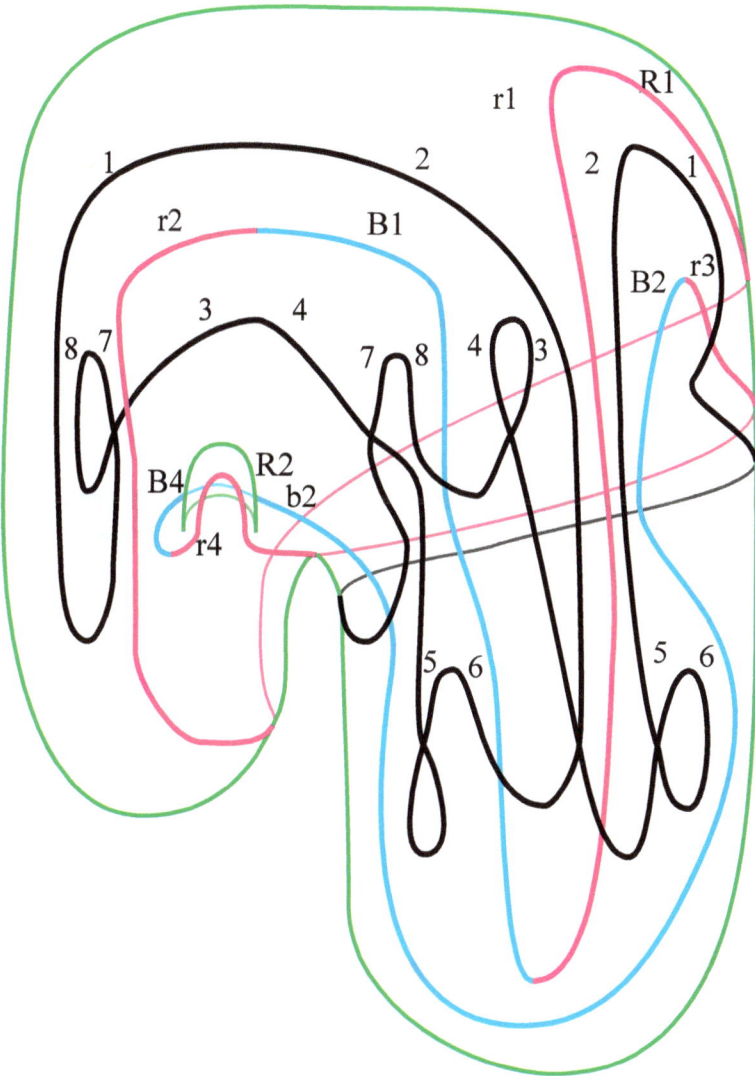

This is illustration blue # 36. In the next stage, a beak-to-beak move cancels the pair of cusps on the back of the figure.

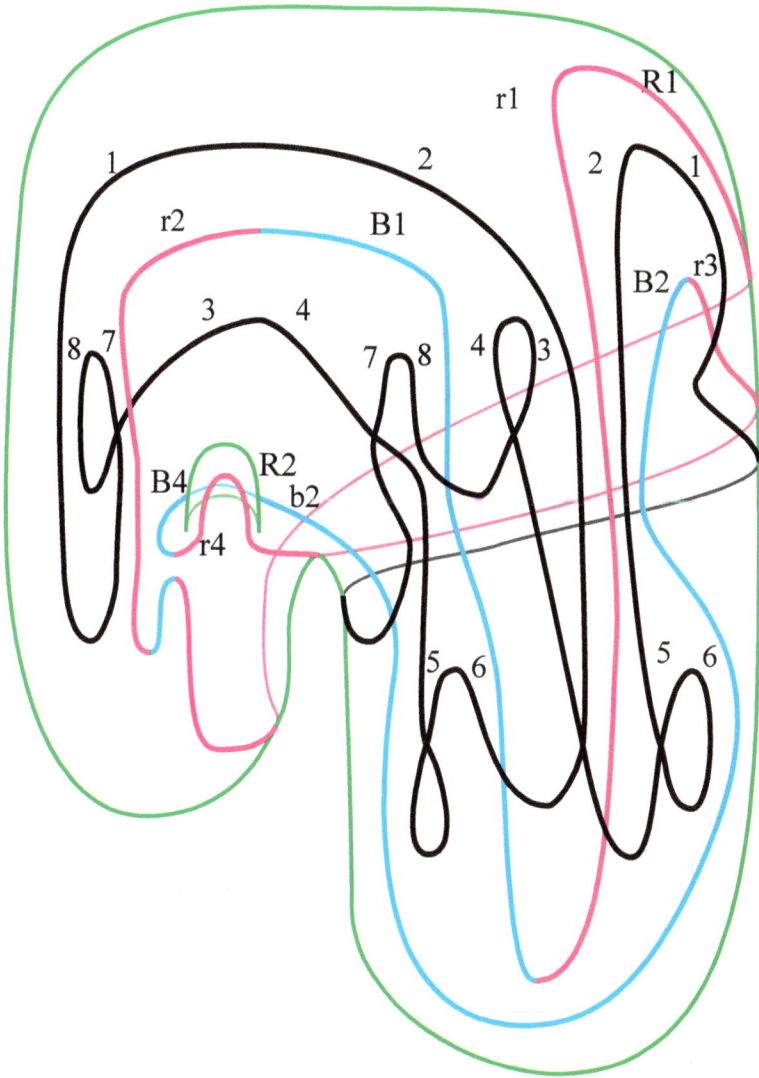

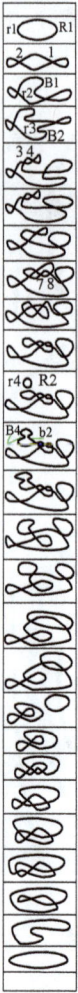

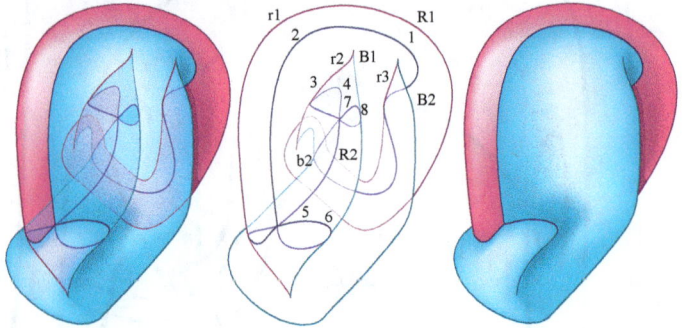

This is illustration blue # 35. In the next stage, the ψ-move on the upper right between double point arc 1 and fold $r3$ passes through the cusp between $r3$ and $B2$ to become a ψ-move between $B2$ and 1. The local red maximum in back cancels with the saddle between $R2$ and $r3$, and the fold $r3$ moves to the far left.

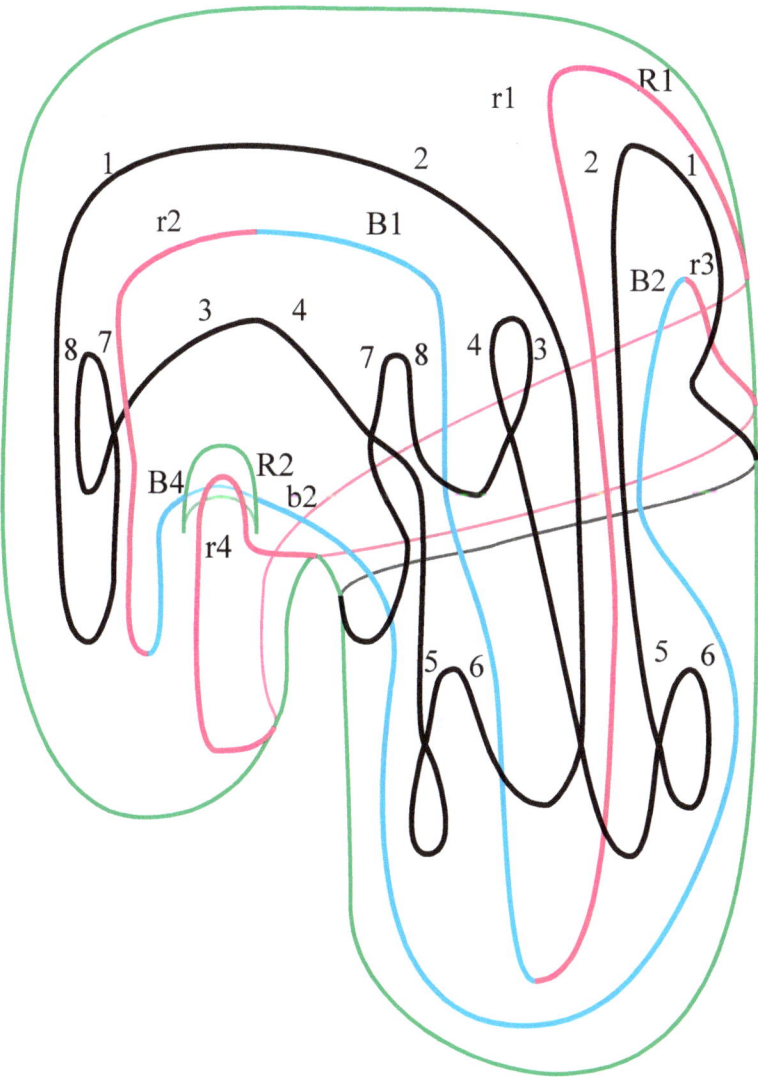

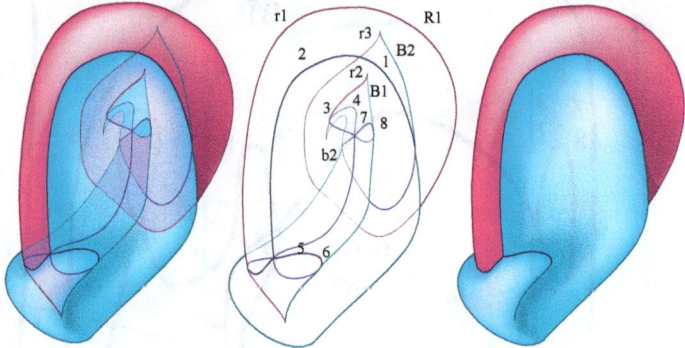

This is illustration blue # 34. In the next stage, the following sequence occurs: a ψ, ψ-move occurs between the fold $b2$ and double point arc 4, the triple point 348 bounces over the birth of 38 and moves behind the fold $b2$, and the top bounce between 4 and $b2$ moves over the saddle death of $B3/b2$. Some vertical positions interchange height.

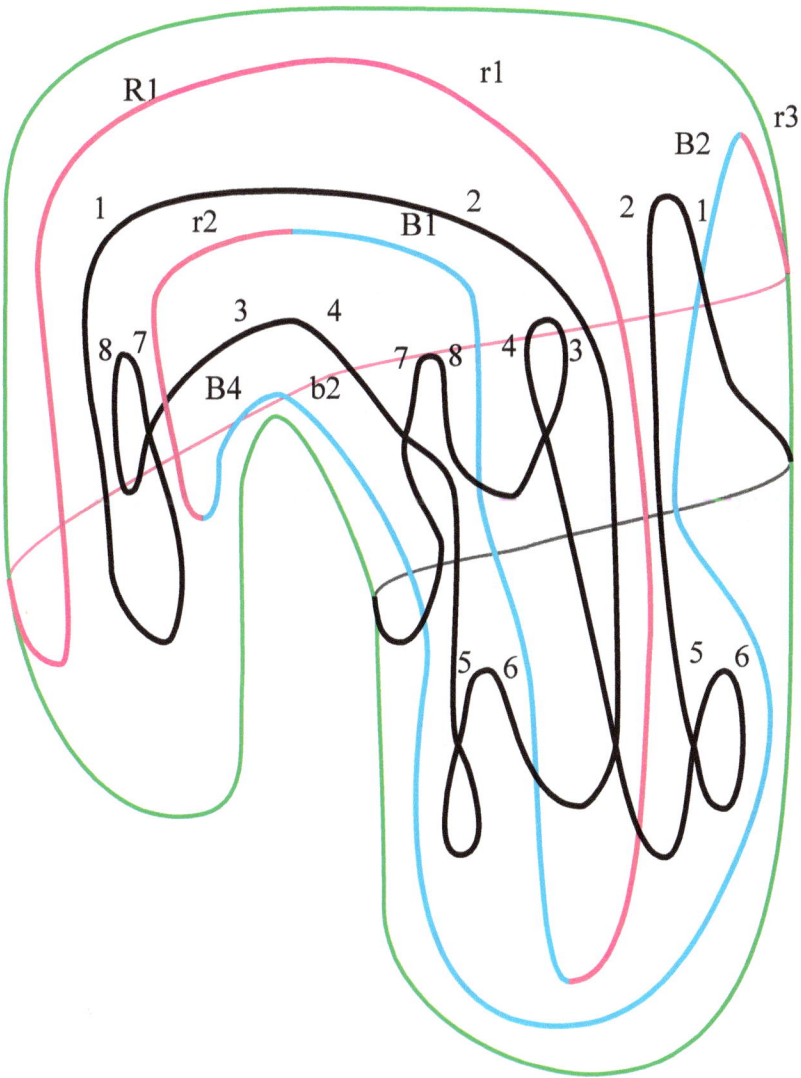

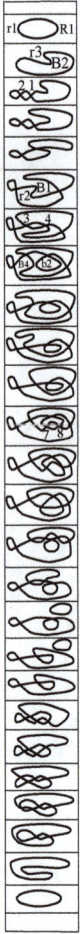

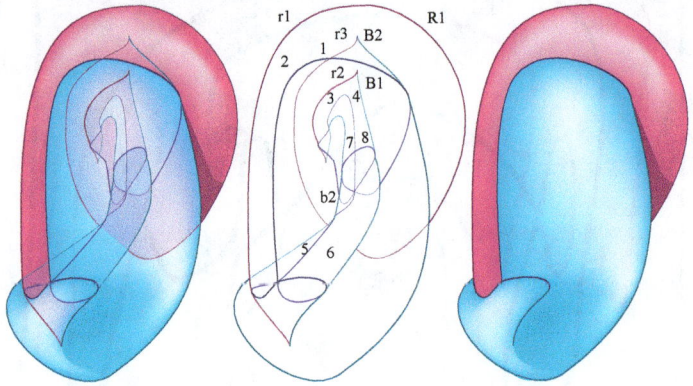

This is illustration blue # 33. In the next stage, the ψ-move between arc 3 and fold $r2$ moves through the cusp $B4/r2$ to become a bounce between $B4$ and double point arc 4.

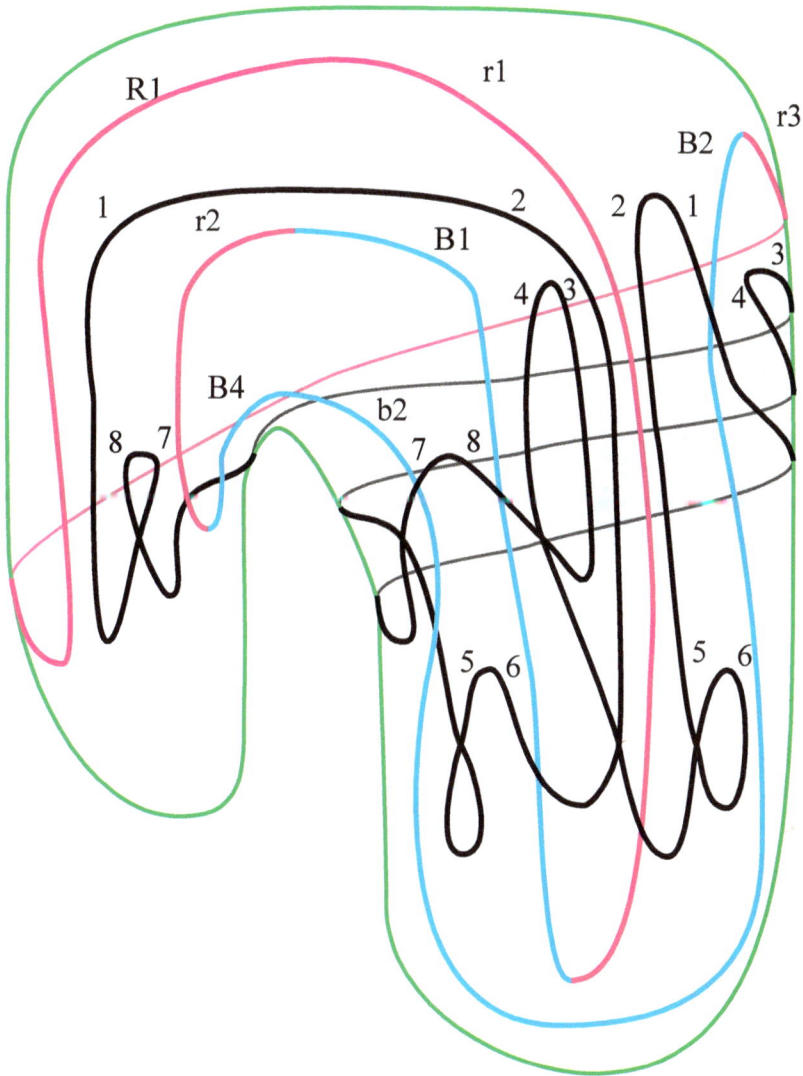

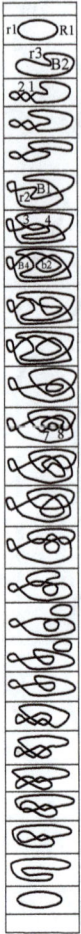

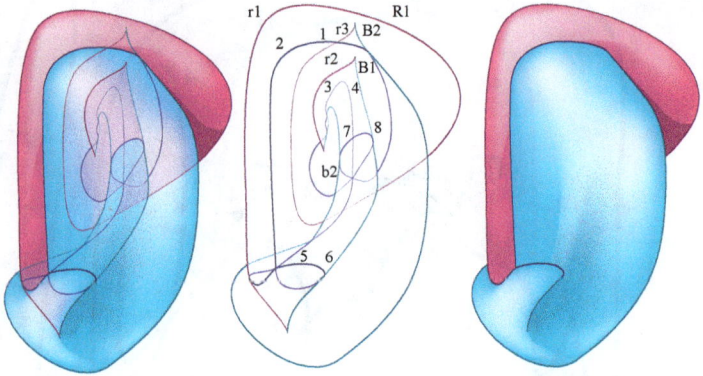

This is illustration blue # 32. In the next stage, the pair of ψ-moves between $B4$ and double point arc 4 are canceled. The death of 34 moves above the $B1/r2$ cusp.

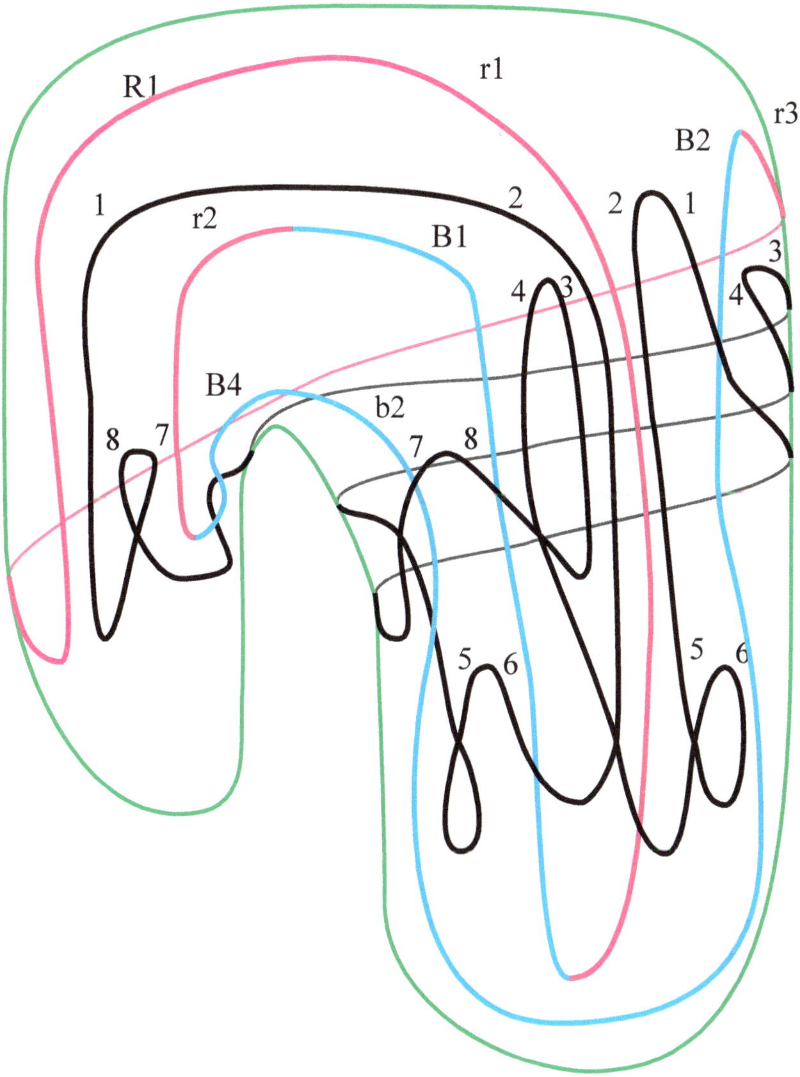
R1
r1
r3
B2
1
2
2
1
r2
B1
3
4
3
4
B4
b2
8
7
7
8
5
6
5
6

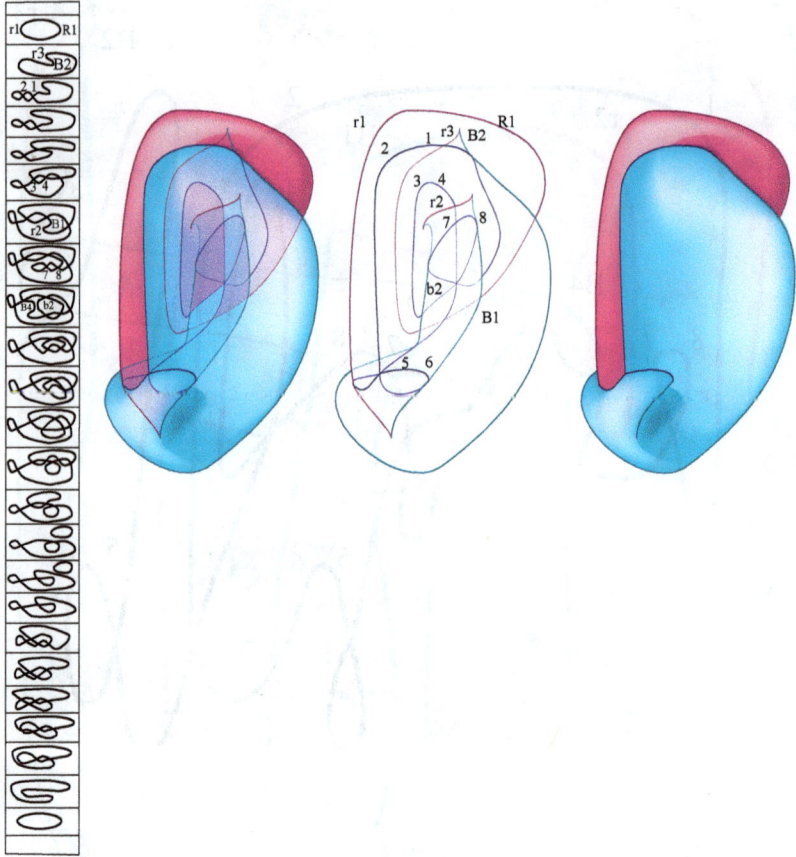

This is illustration blue # 31. In the next stage, a type II saddle move occurs between the double point arcs 1 and 4.

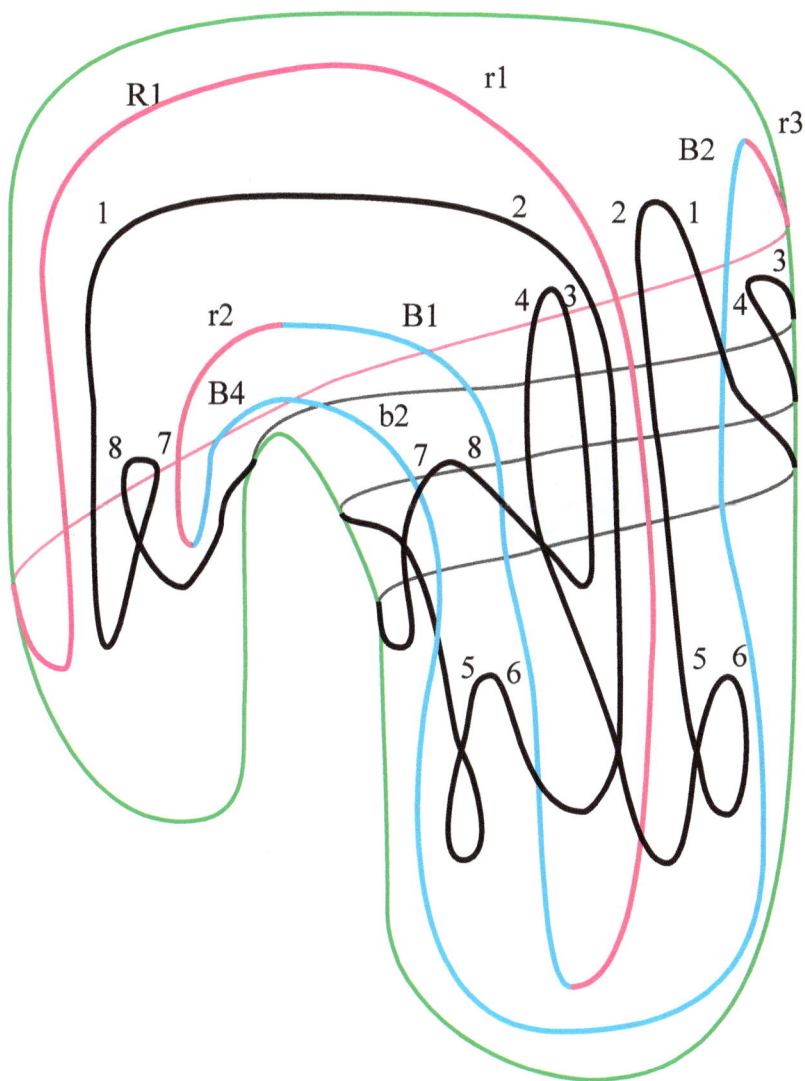

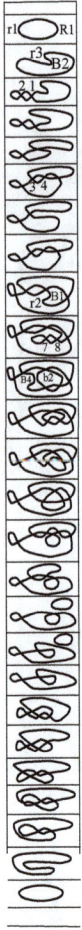

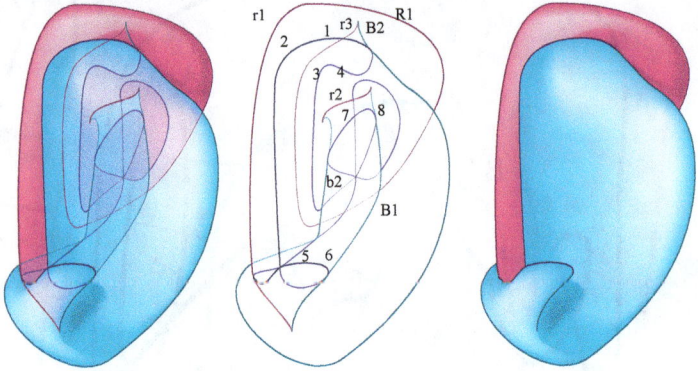

This is illustration blue # 30. In the next stage, a type II zig-zag move cancels the vestigial death of 34 and the bounce between 1 and $r3$ moves over the cuspidal birth of $r3$ and $B2$. The remaining death is renumbered.

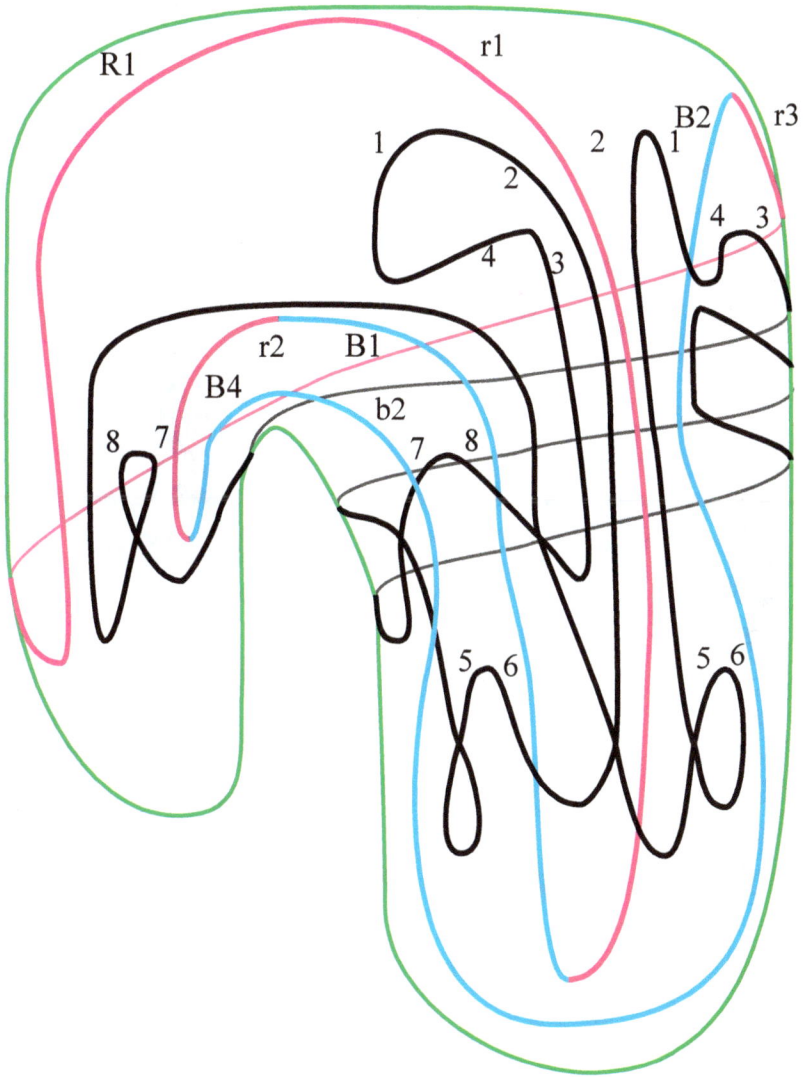

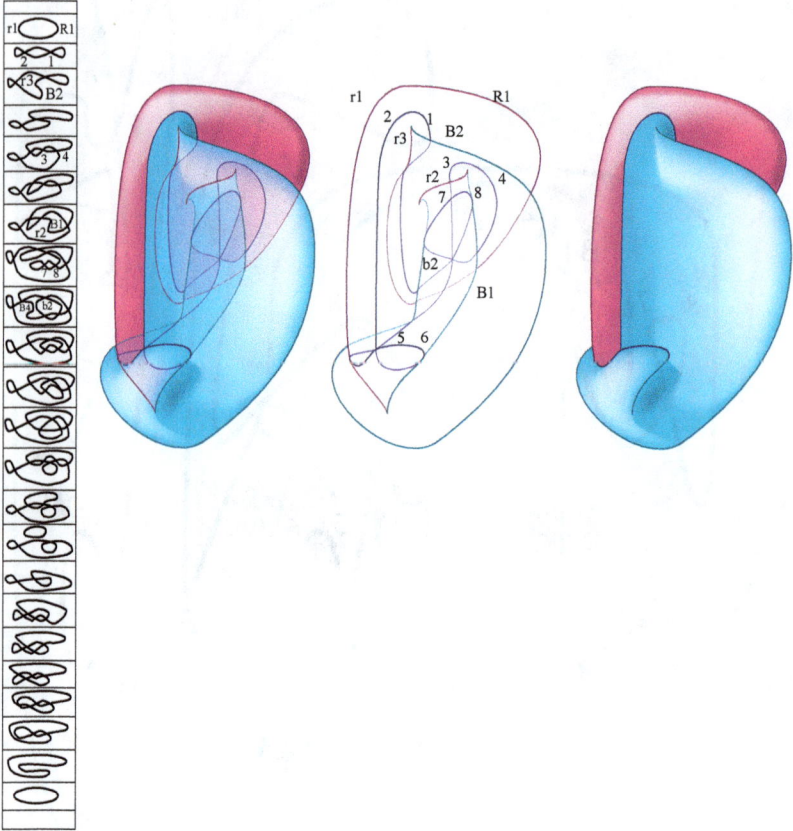

This is illustration blue # 29. In the next stage, a canceling pair of death and saddle occurs along on the blue curve $B2$ and the ϕ-move between double point arc 1 and red arc $r3$ moves to the bottom of the figure.

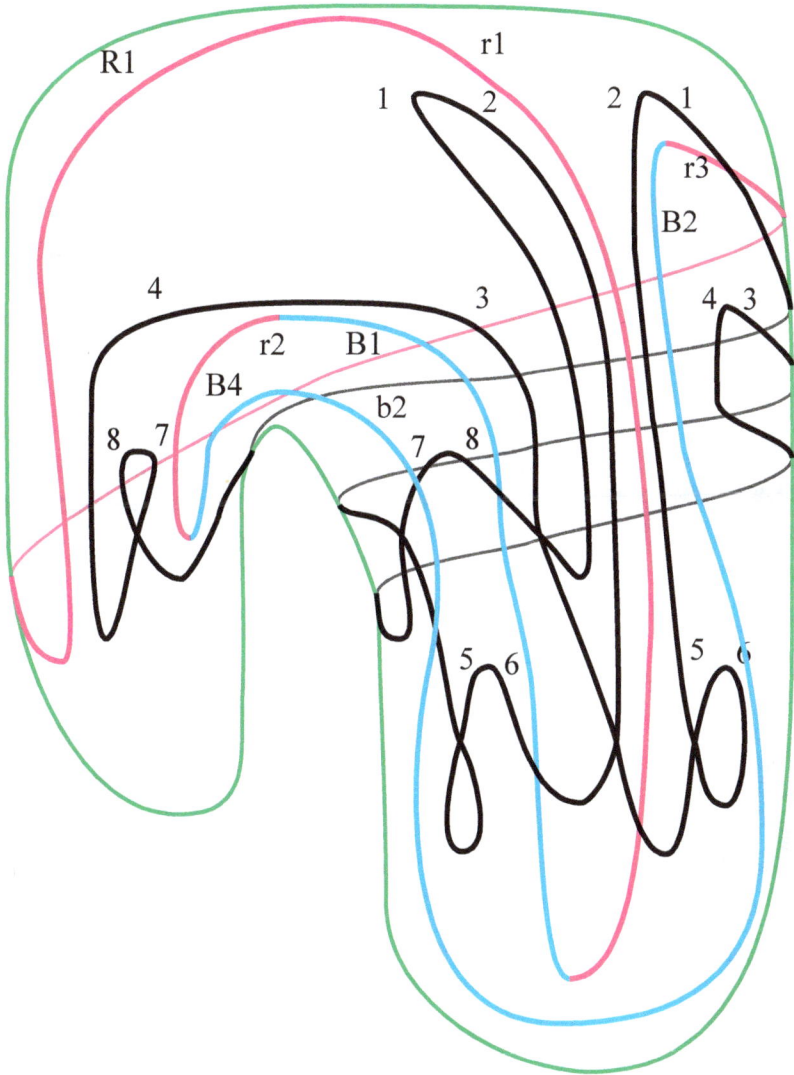

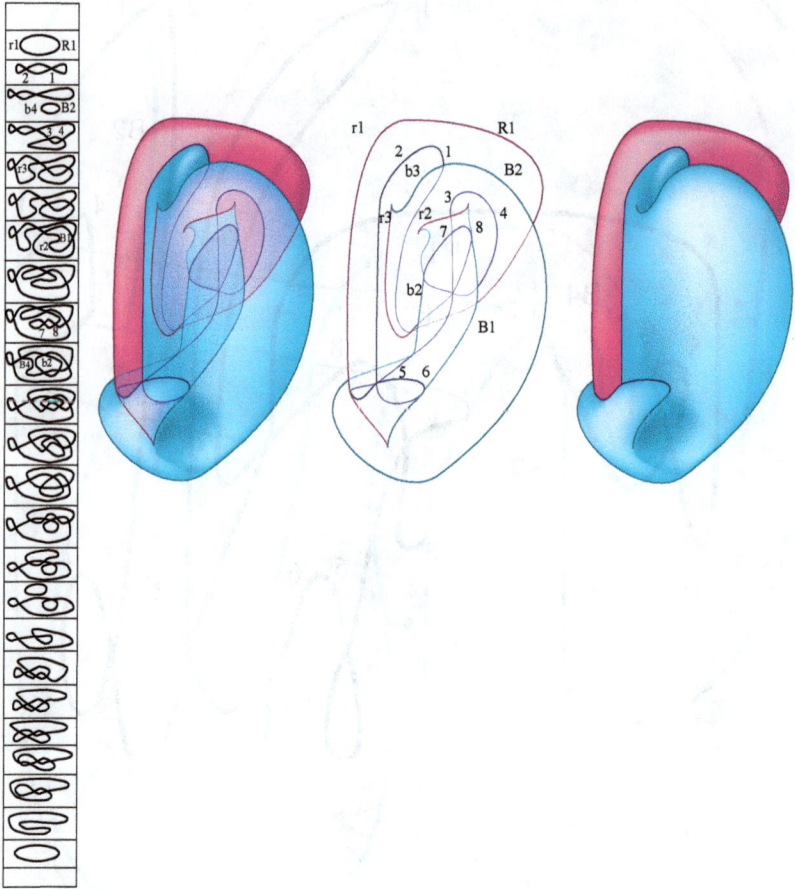

This is illustration blue # 28. In the next stage, a horizontal cusp occurs on the upper left.

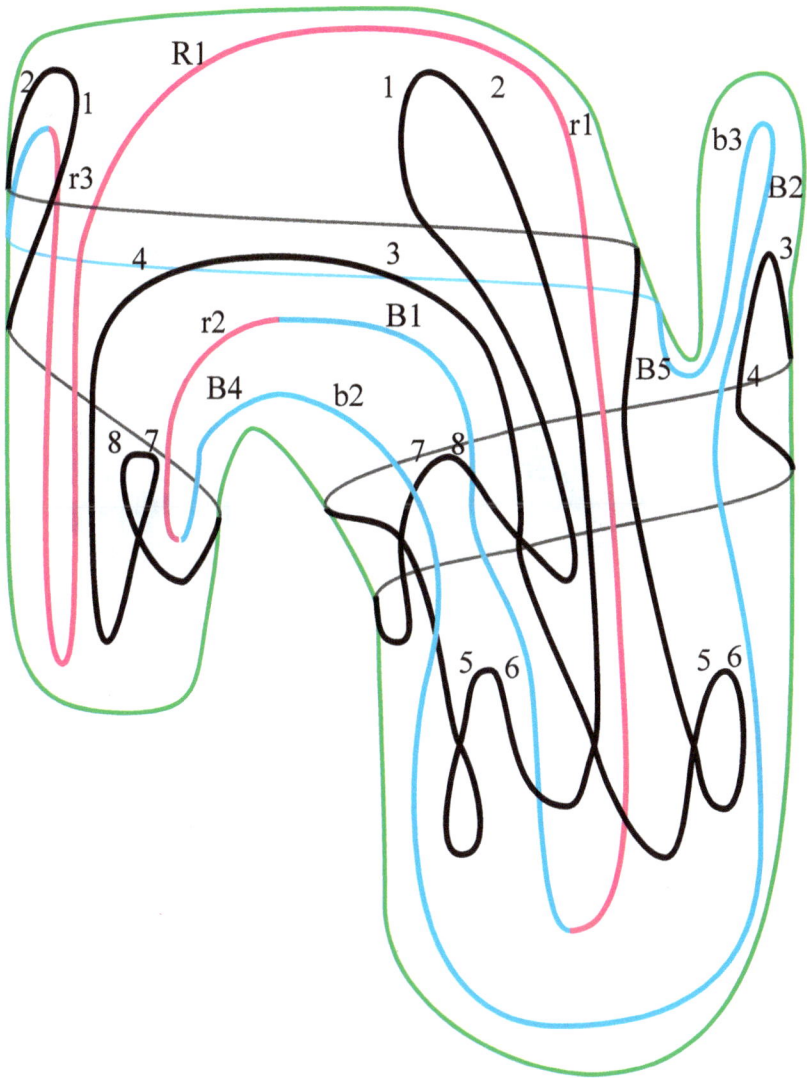

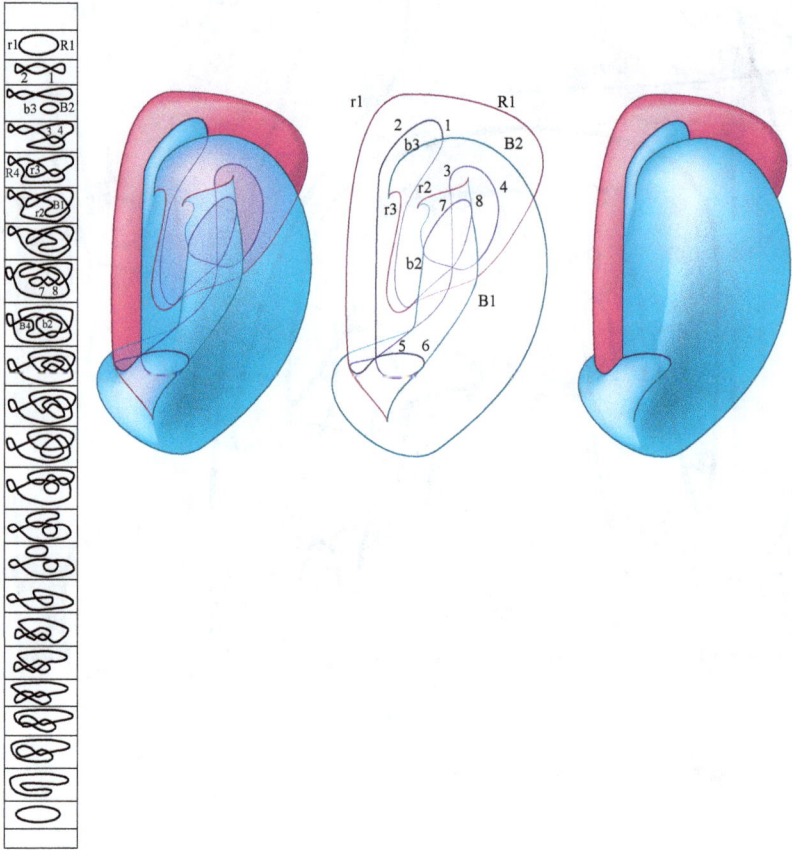

This is illustration blue # 27. In the next stage, the two saddle points interchange heights.

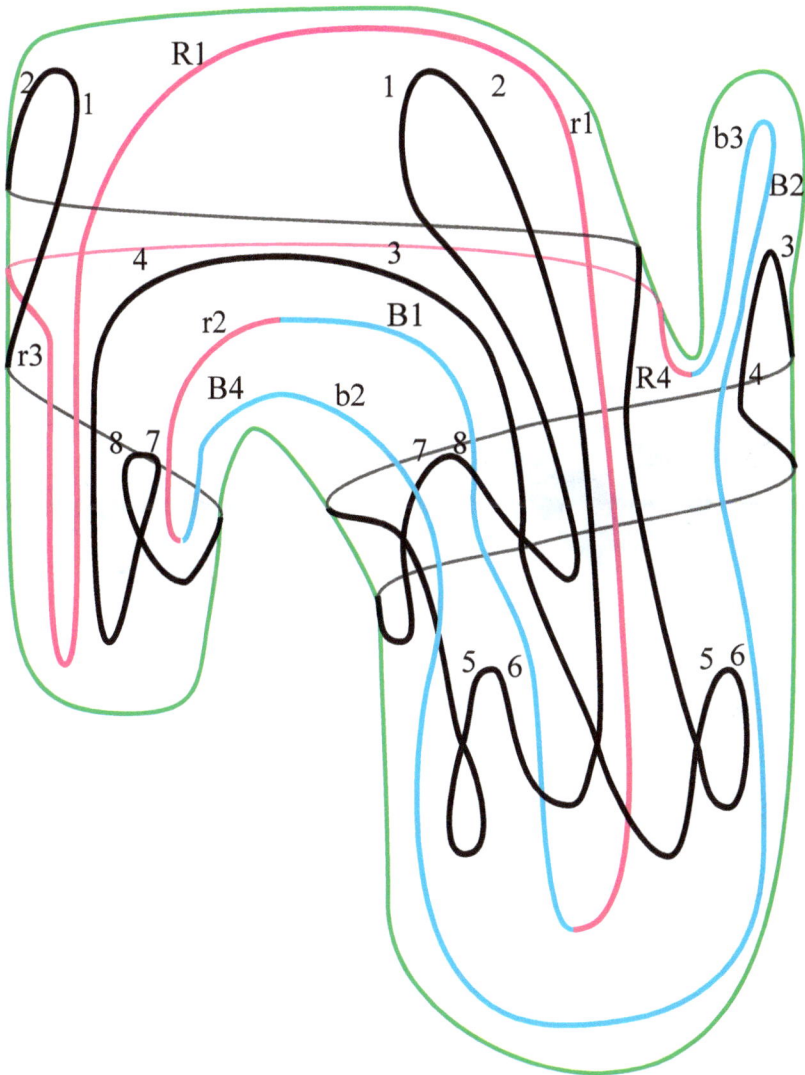

The decker set in the next illustration, is a quite complicated deformation of that in this illustration. In fact, I spent more time on these two decker sets than every other decker illustration. It takes a great deal of patience to see that there is a deformation between them.

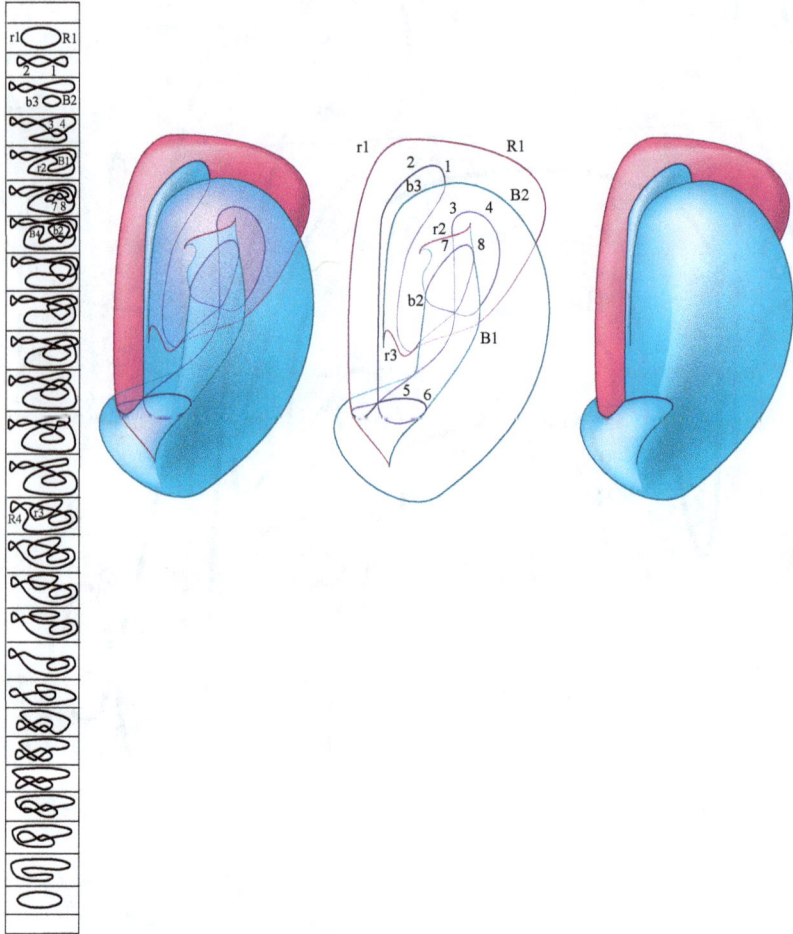

This is illustration blue # 26. In the next stage, the bounce between the birth of 18 and the arc $r3$ passes over the saddle. Then there is a critical cancellations between the birth of $r3 - R1$ and the saddle $R4 - r3$.

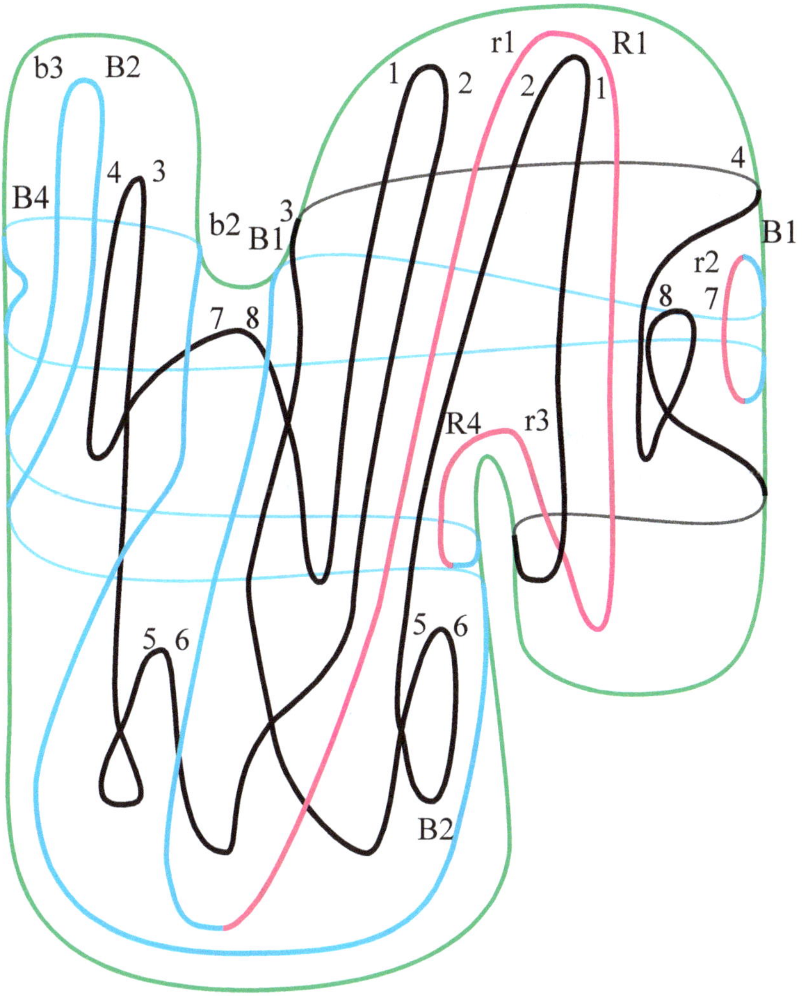
b3
B2
4
3
B4
b2
B1
3
1
2
r1
R1
2
1
4
B1
r2
8
7
7
8
R4
r3
5
6
5
6
B2

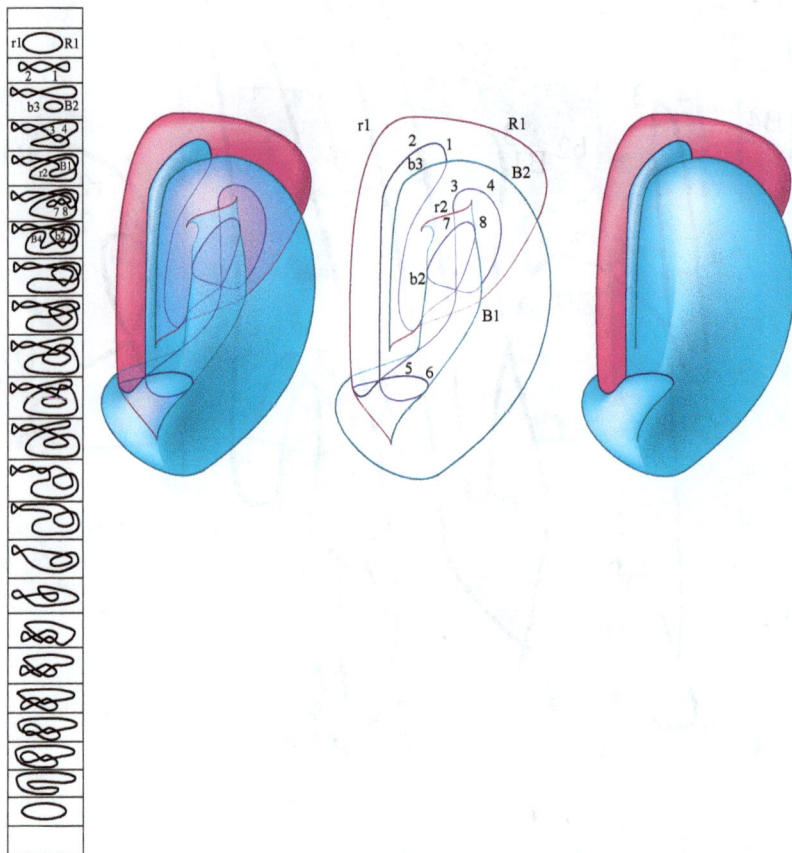

This is illustration blue # 25. In the next stage, the birth of double points 1 and 8 cancel, via a type II zig-zag move with the death of 1 and 2.

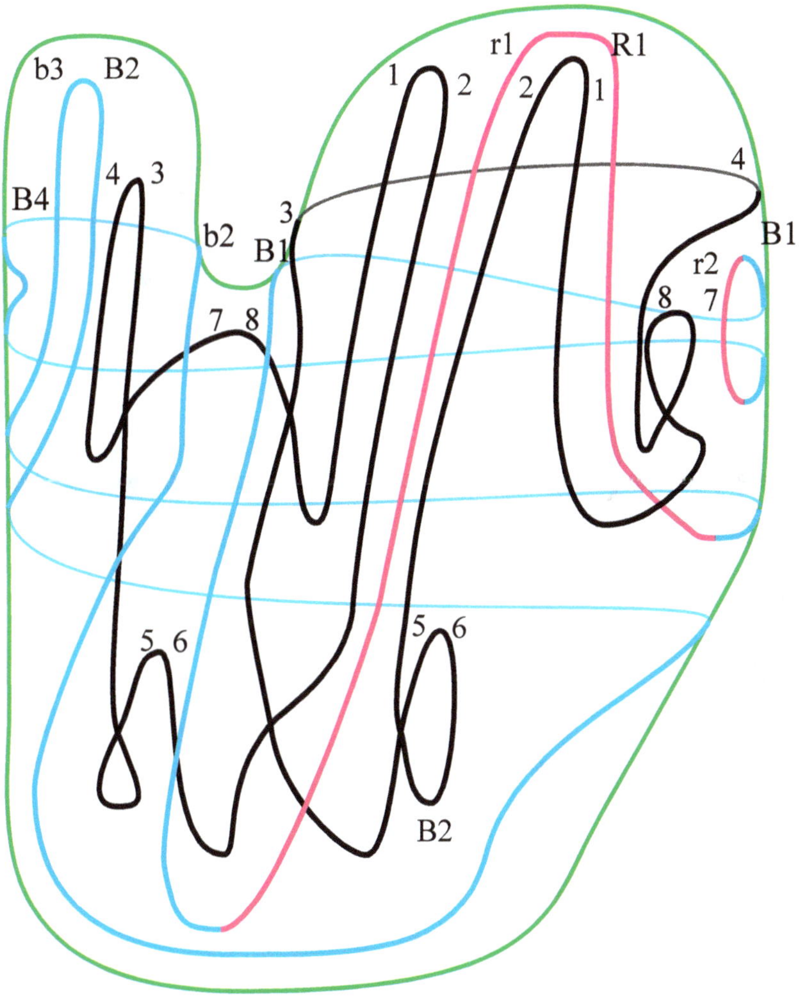
b3
B2
4
3
B4
b2
3
B1
7
8
1
2
r1
R1
2
1
4
B1
r2
8
7
5
6
5
6
B2

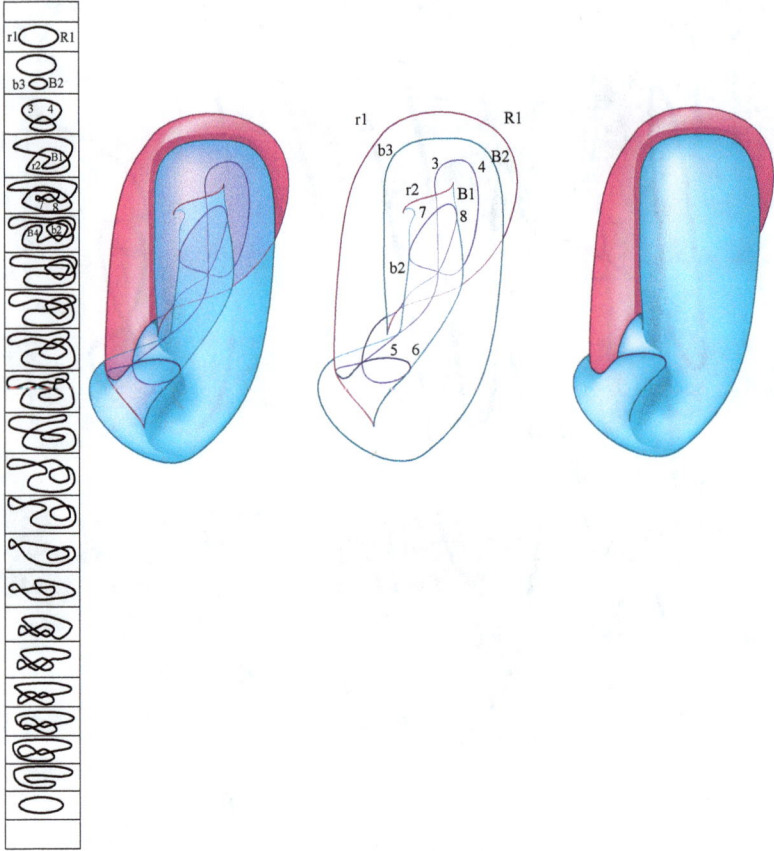

This is illustration blue # 24. In the next stage, blue fold $b2$ peeks out from behind fold $b3$ on the left and the birth of double point arc 47 bounces to the left of the triple point.

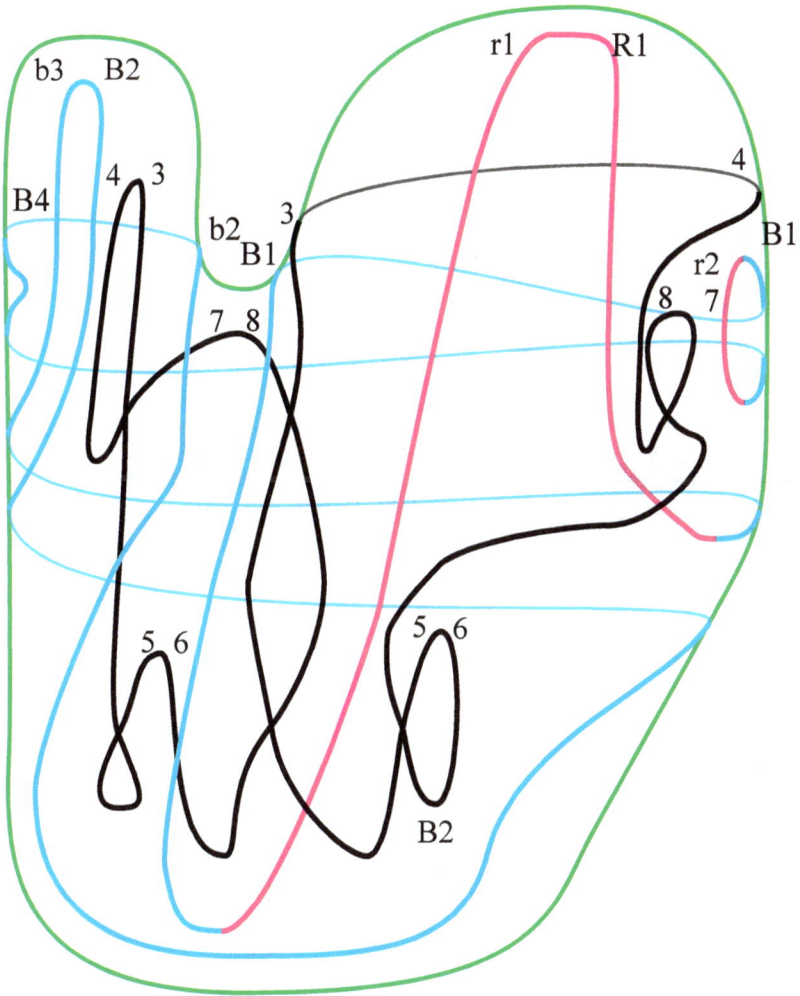

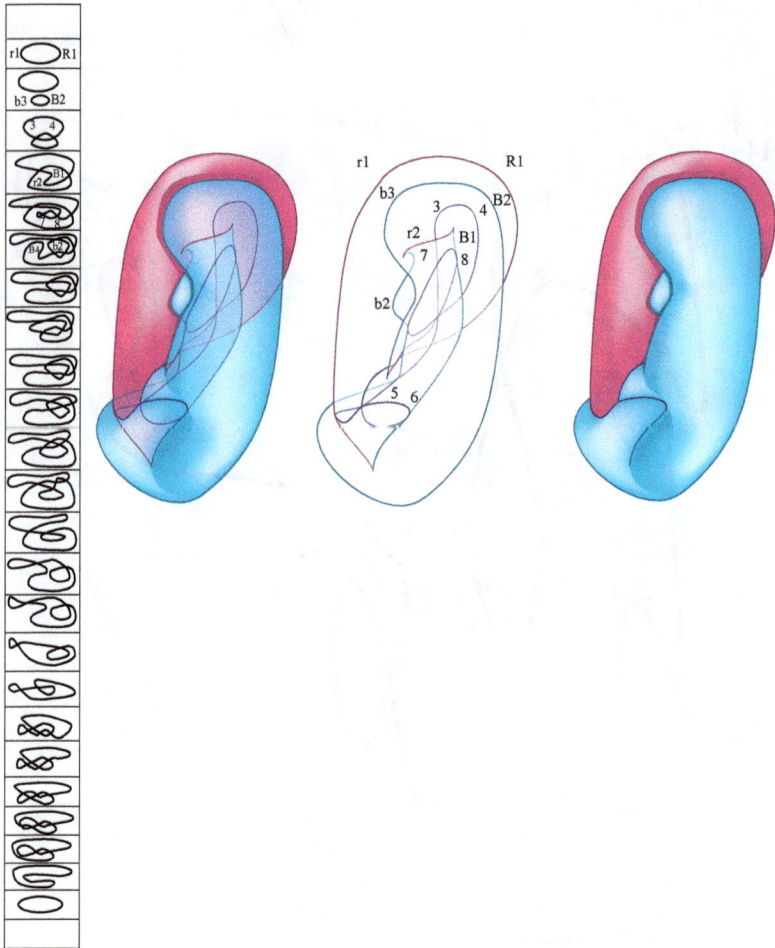

This is illustration blue # 23. In the next stage, double point arc 8 and fold $b3$ undergo a ψ, ψ-move.

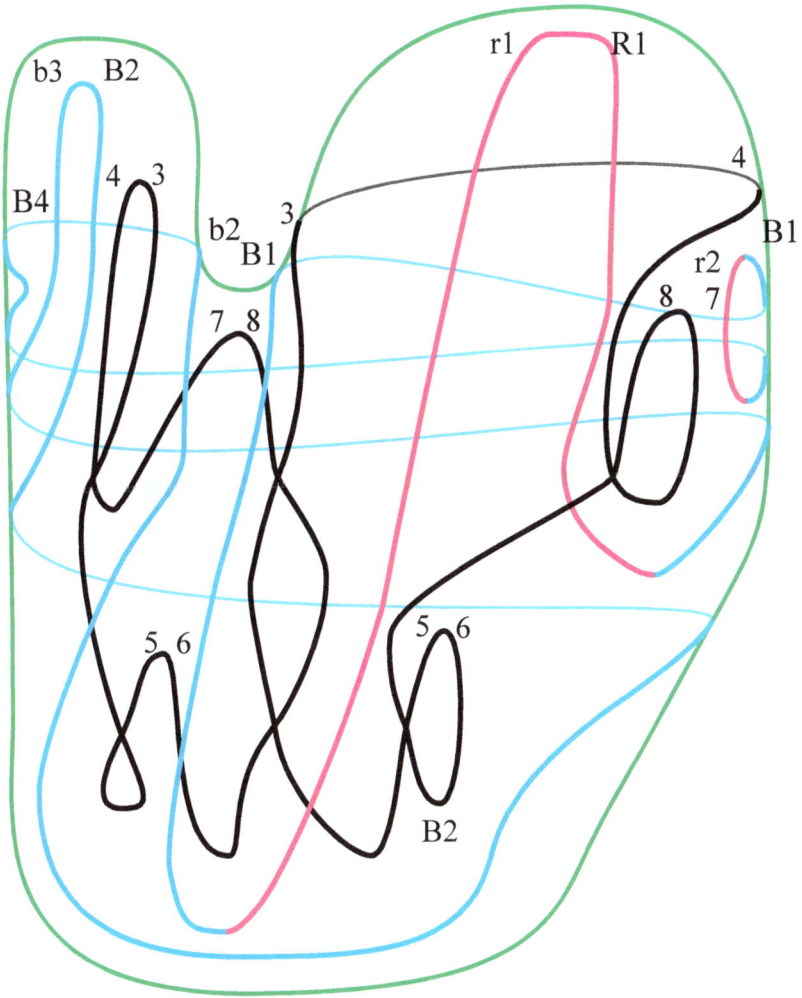

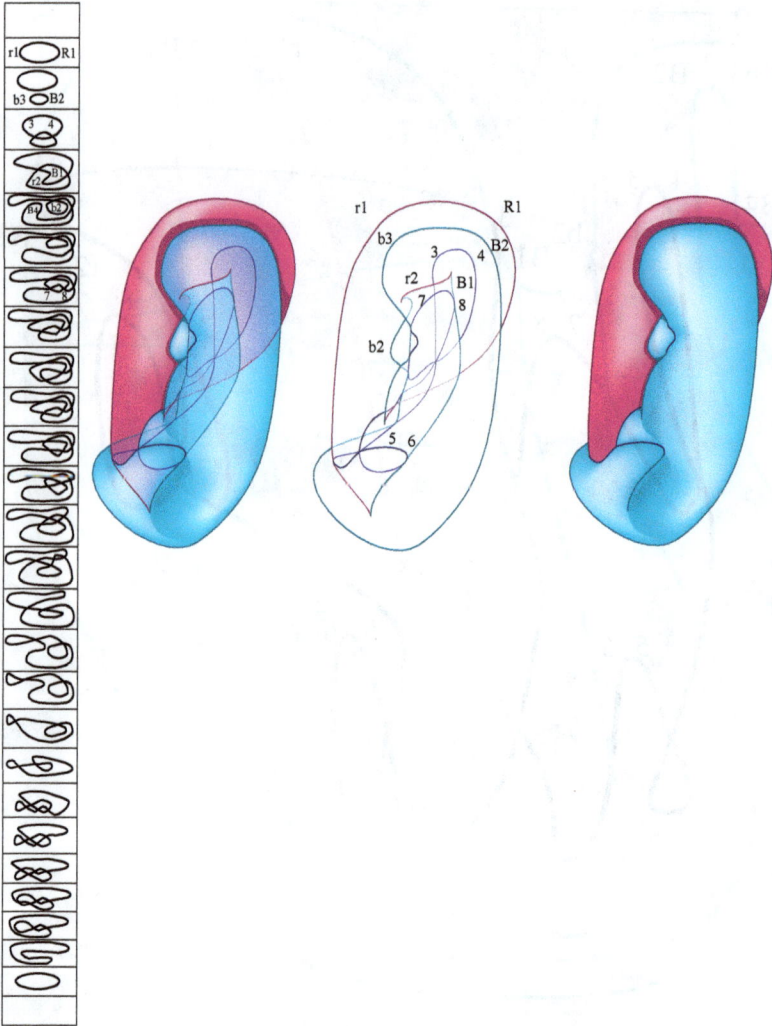

This is illustration blue # 22. In the next stage, the death of double arcs 5 and 6 moves up while the double point arc 7 acquires a zig-zag bend (type II zig-zag) with a birth and arc labeled 0 and a relabeling of arc 7 by the number 9.

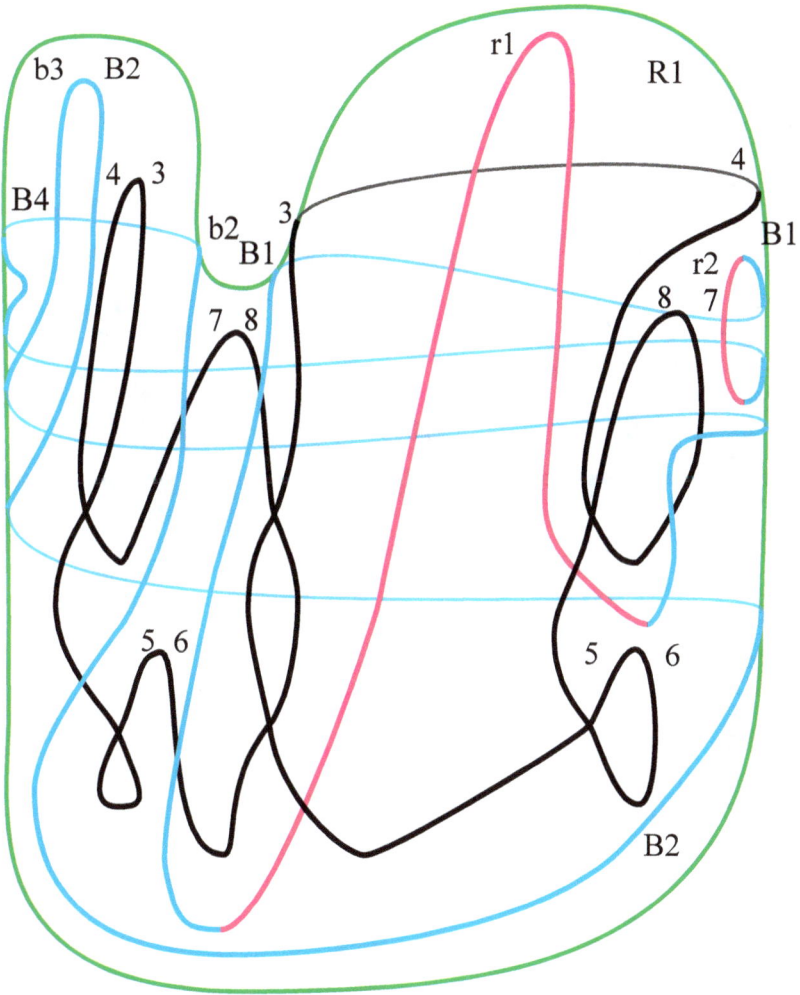

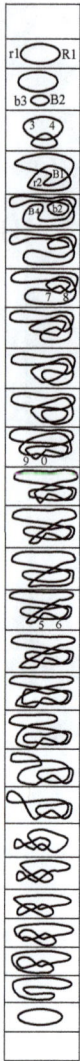

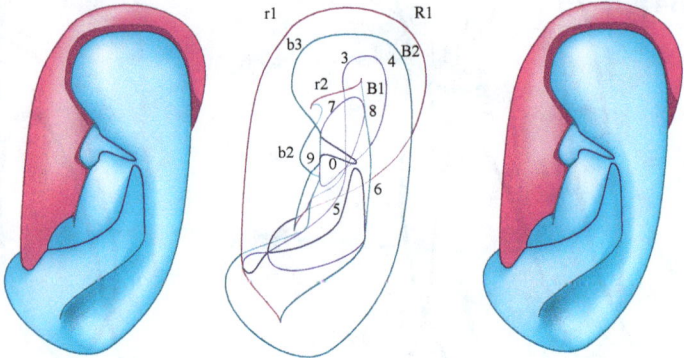

This is illustration blue # 21. In the next stage, the death of double arcs 5 and 6 cancels with the birth of 07 by a type II saddle move.

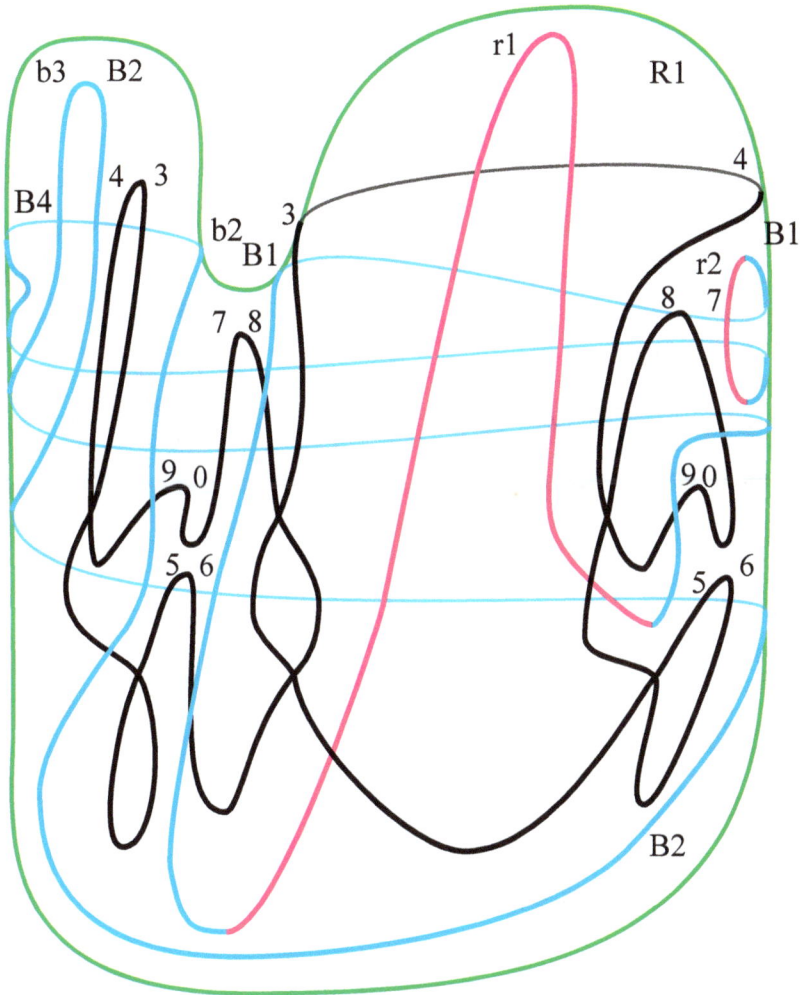

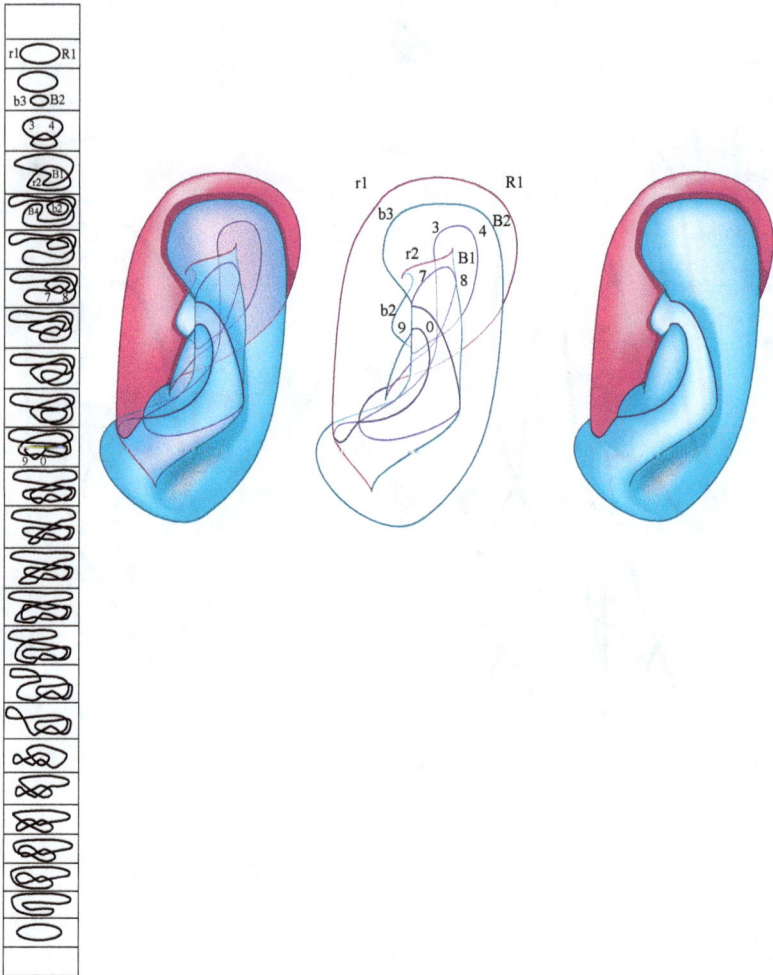

This is illustration blue # 20. In the next stage, a horizontal type II move occurs and type II zig-zig move cancels the spurious death 09 against the birth of 39.

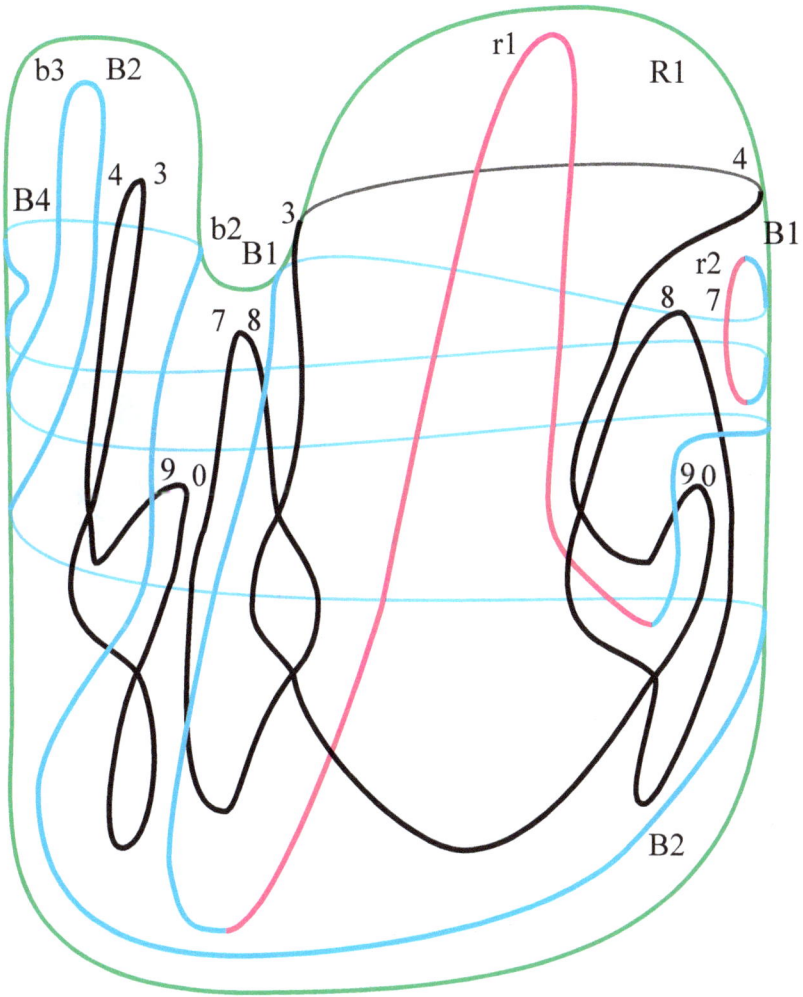

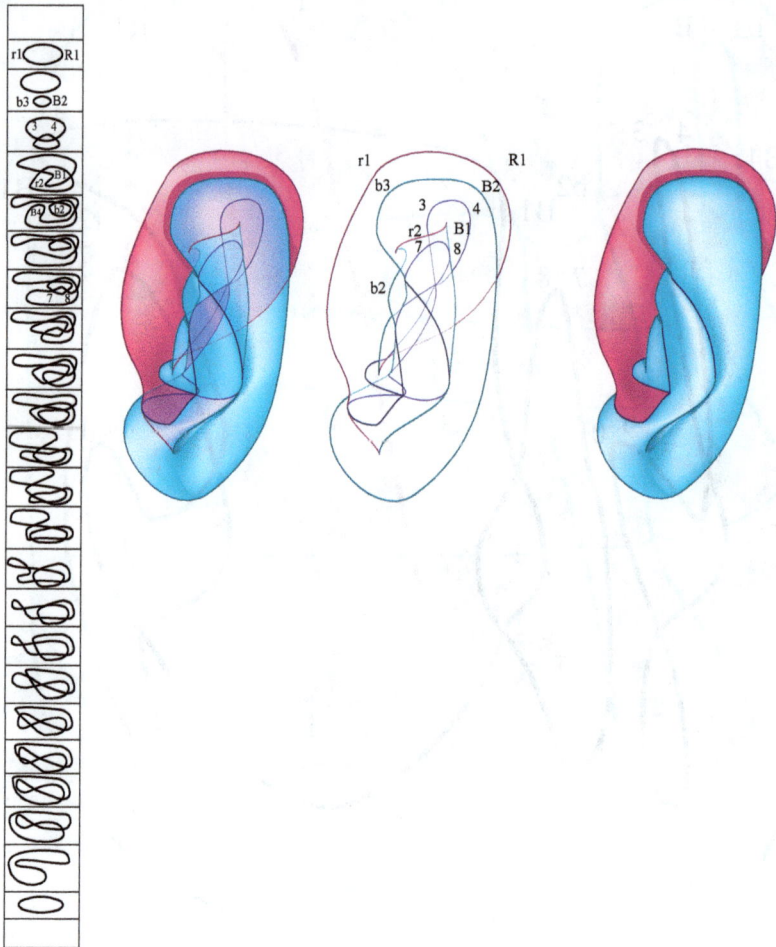

This is illustration blue # 19. In the next stage, the top triple point 348 passes over the fold *b*3, and various heights are readjusted.

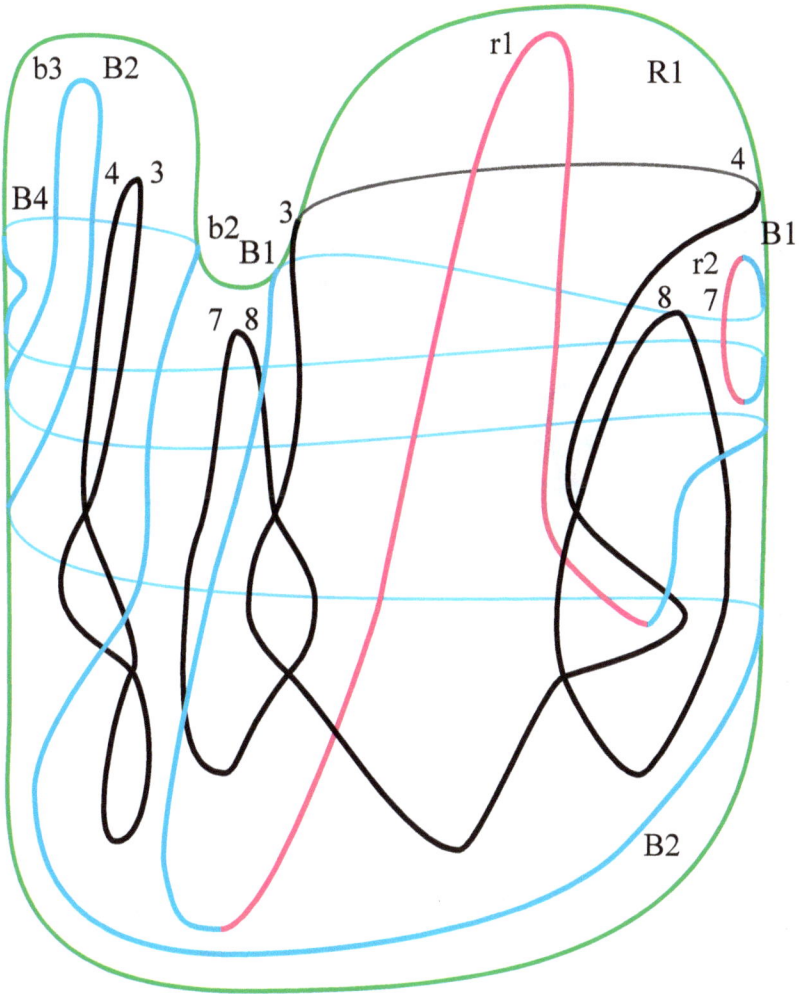

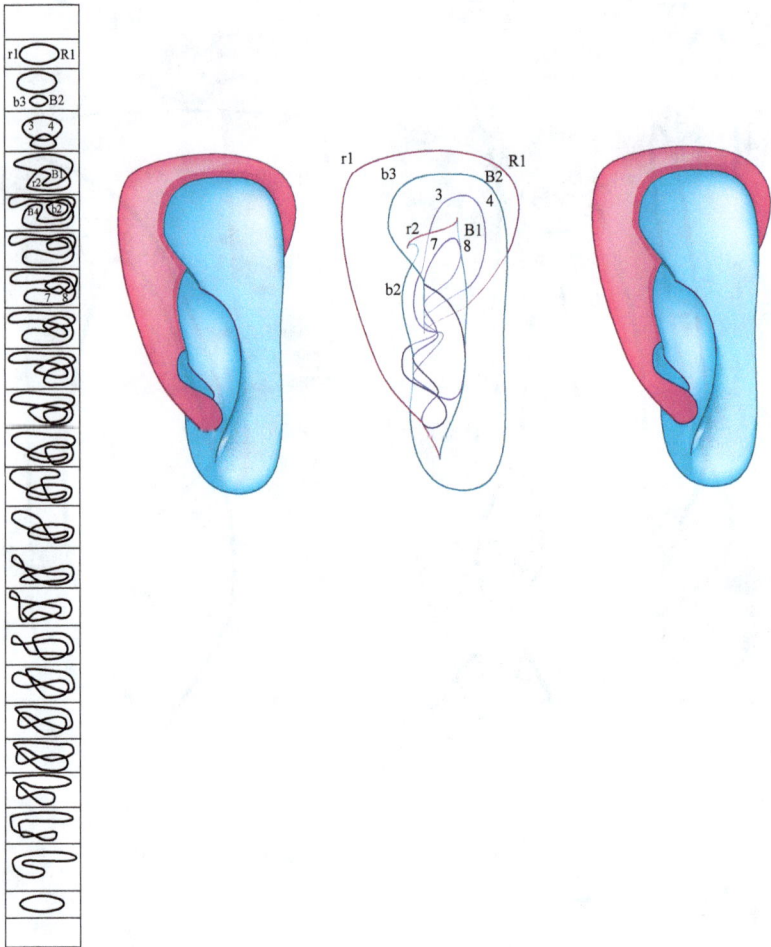

This is illustration blue # 18. In the next stage, the bottom triple point 348 passes to the other side of the fold *b*2.

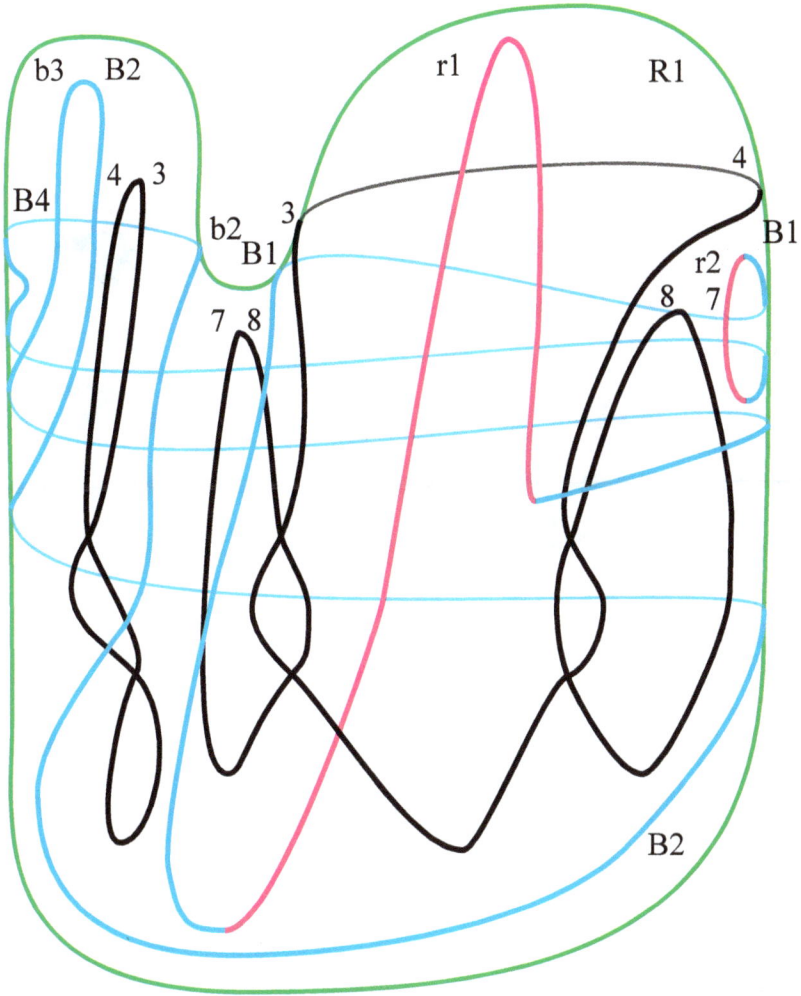

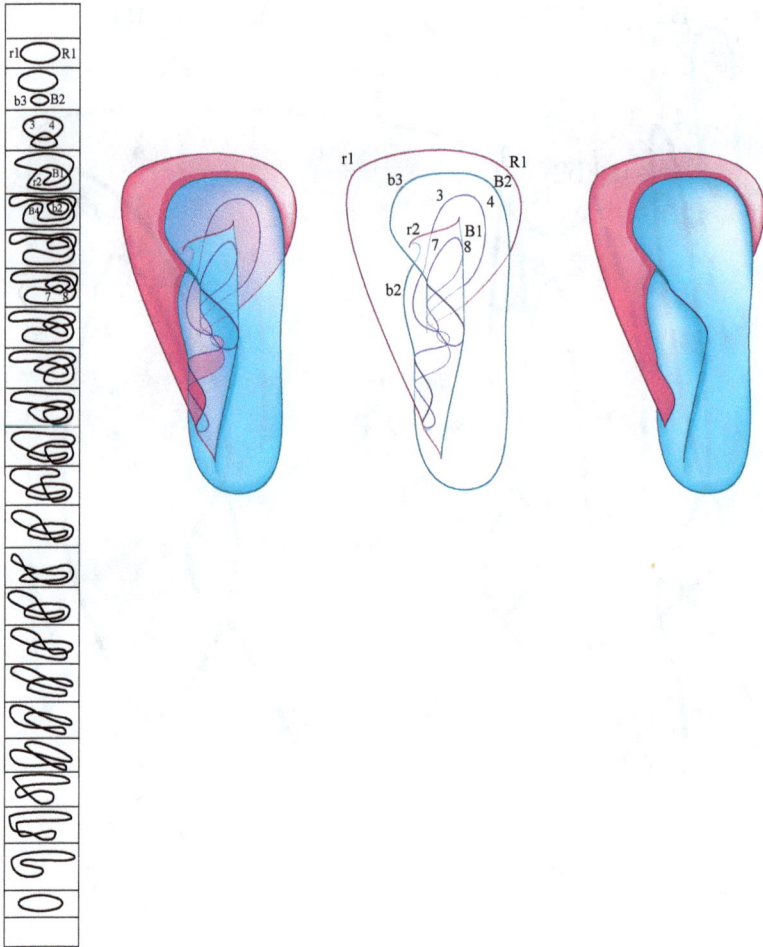

This is illustration blue # 17. In the next stage, a horizontal type II move occurs at the bottom and a ψ, ψ-move cancels the redundant bounces.

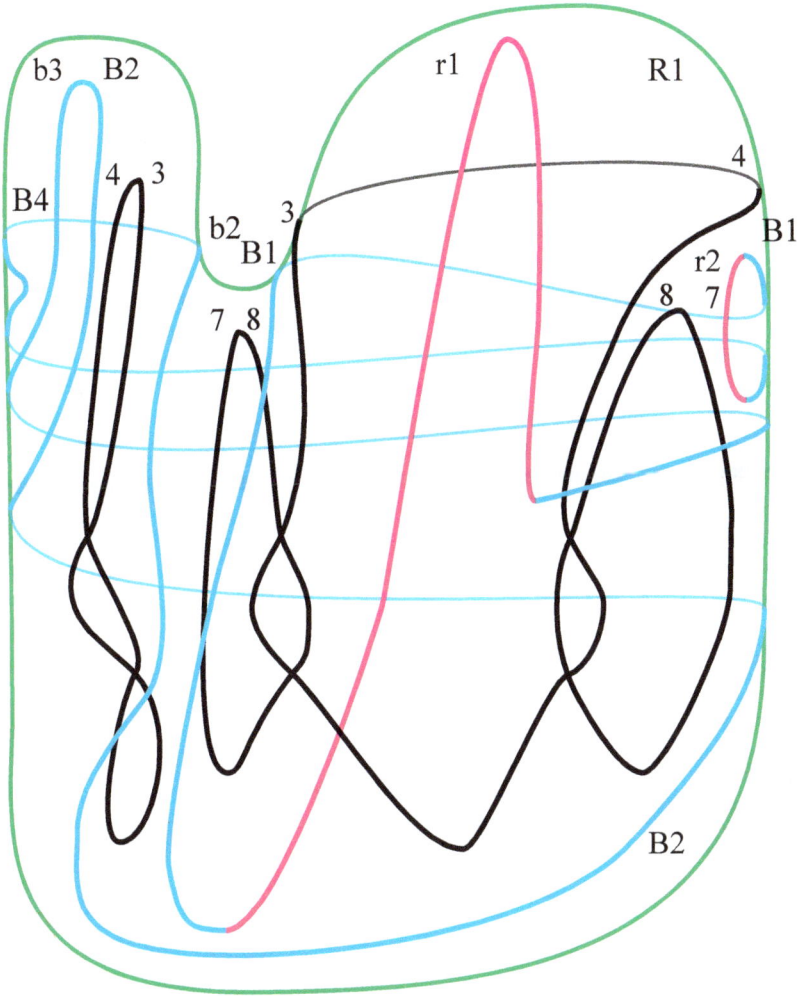

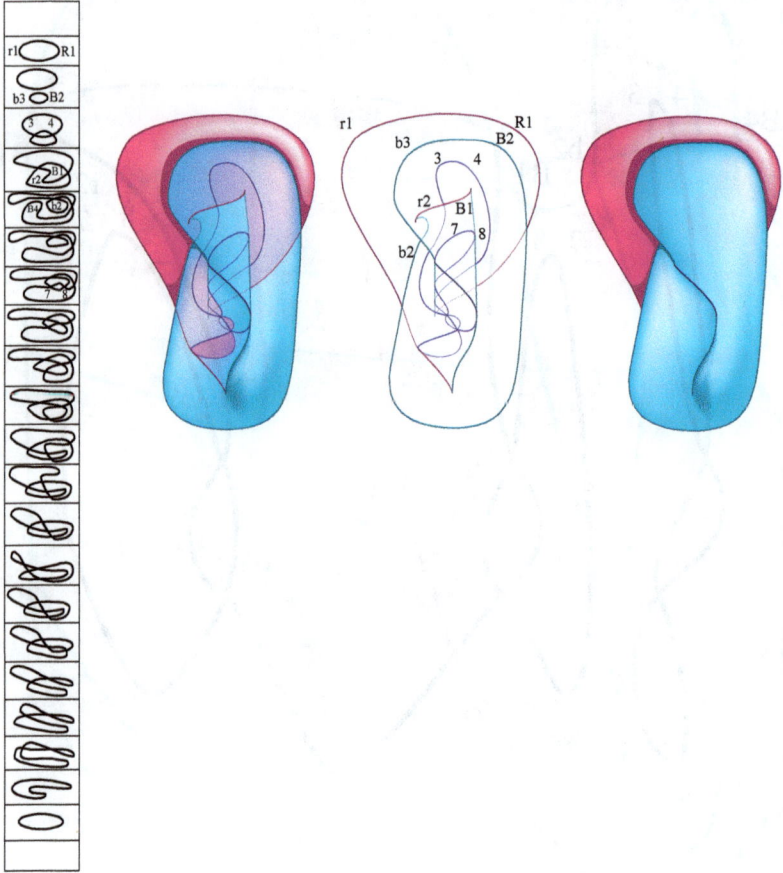

This is illustration blue # 16. In the next stage, the two triple points cancel.

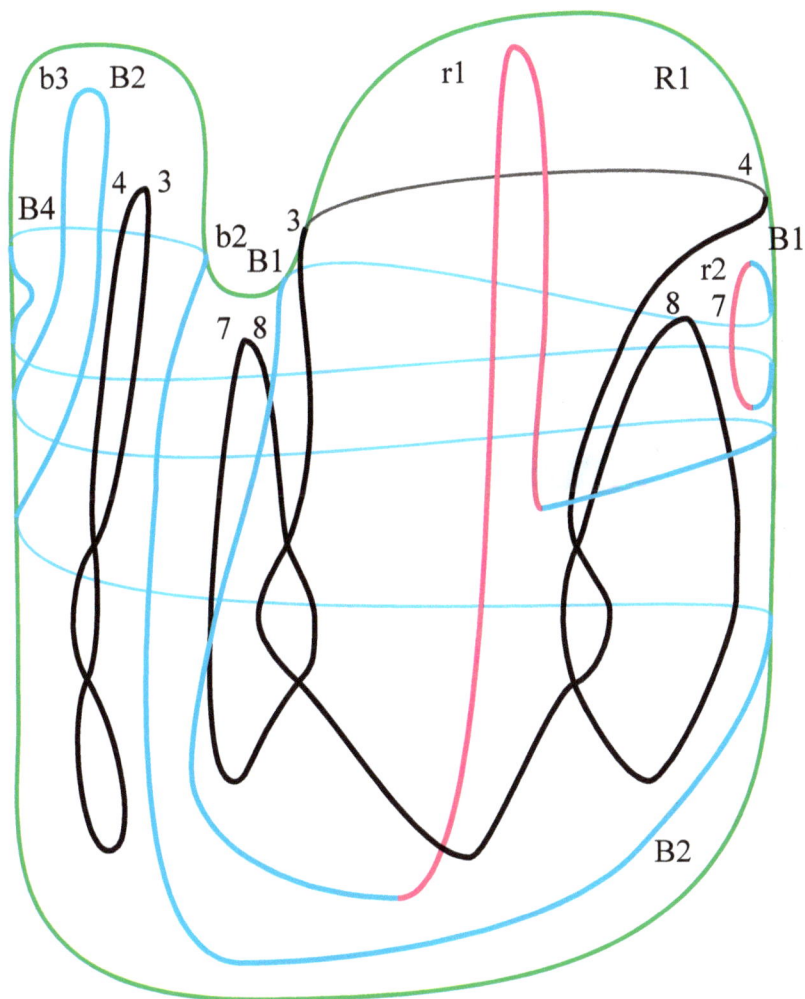

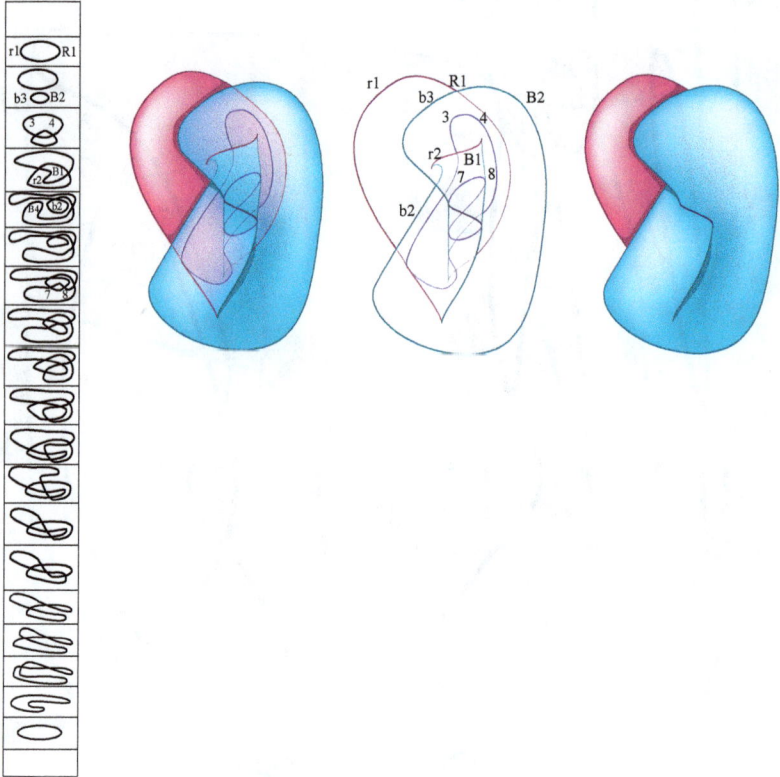

This is illustration blue # 15. In the next stage, double point arc 4 bounces to the other side of the b3 R1 cusp. The cusp moves to the right.

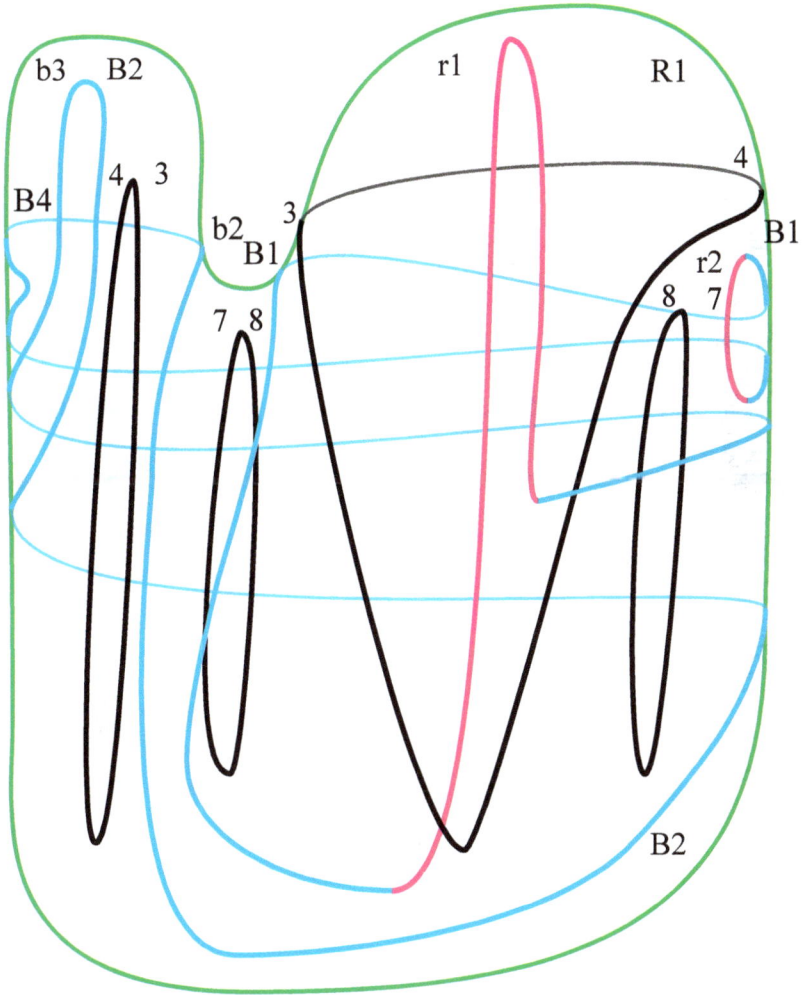

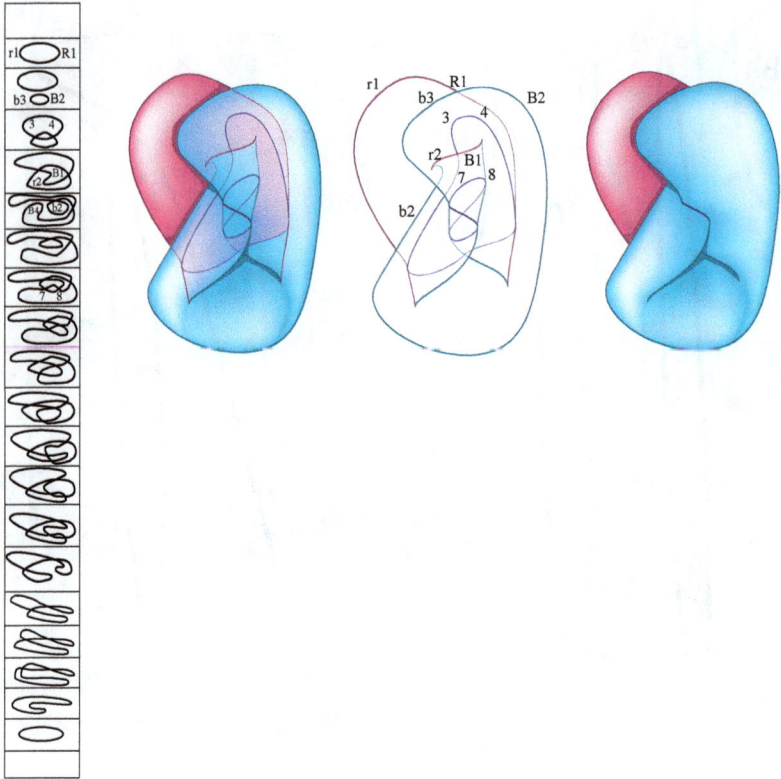

This is illustration blue # 14. In the next stage, a horizontal type II move occurs at the birth of 78 and the birth of 34 becomes visible. Heights are adjusted.

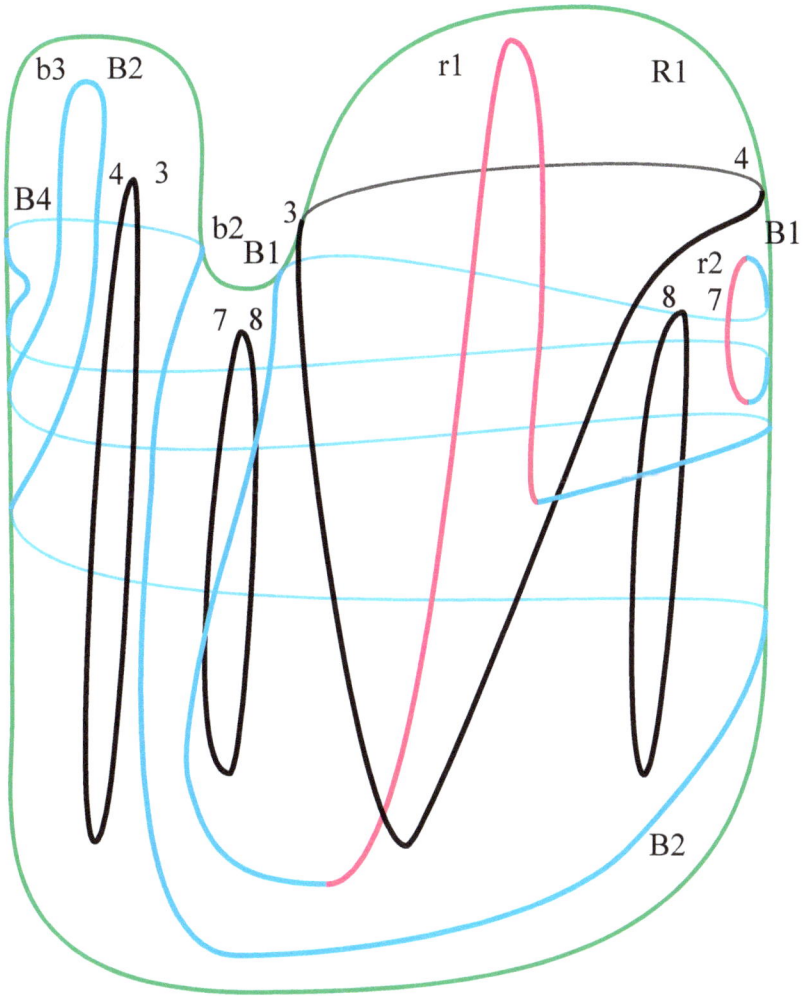

This is illustration blue # 13. In the next stage, two ψ, ψ-moves elimi-
nate the extra bounces between double loop 78 and the folds.

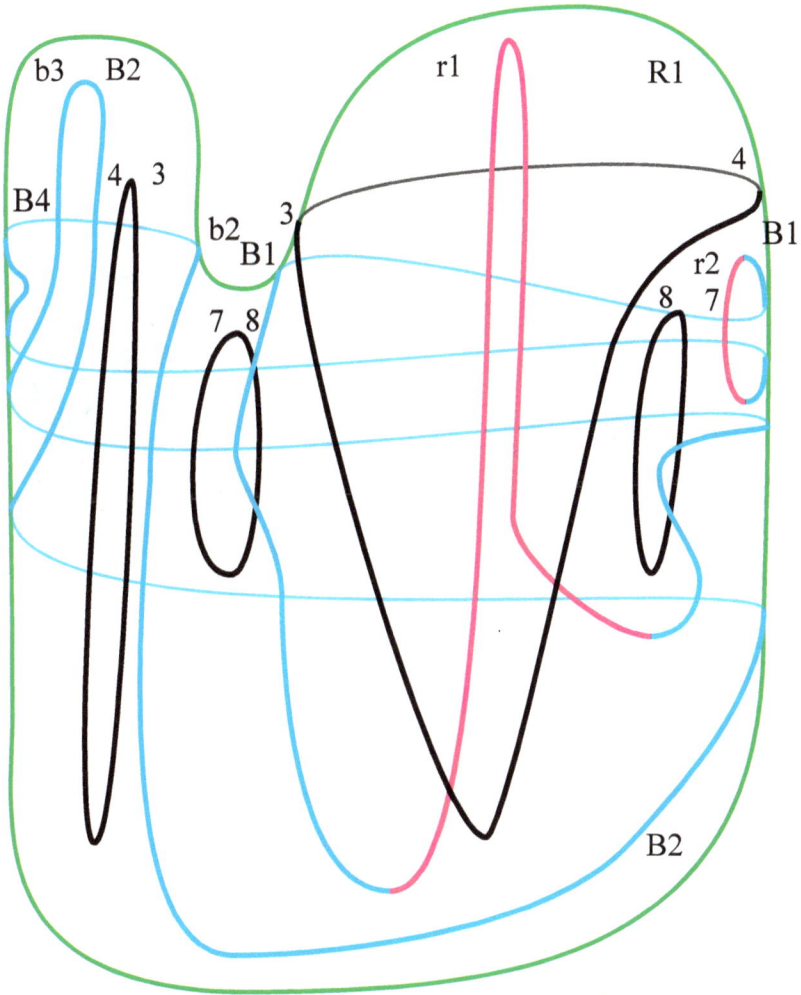

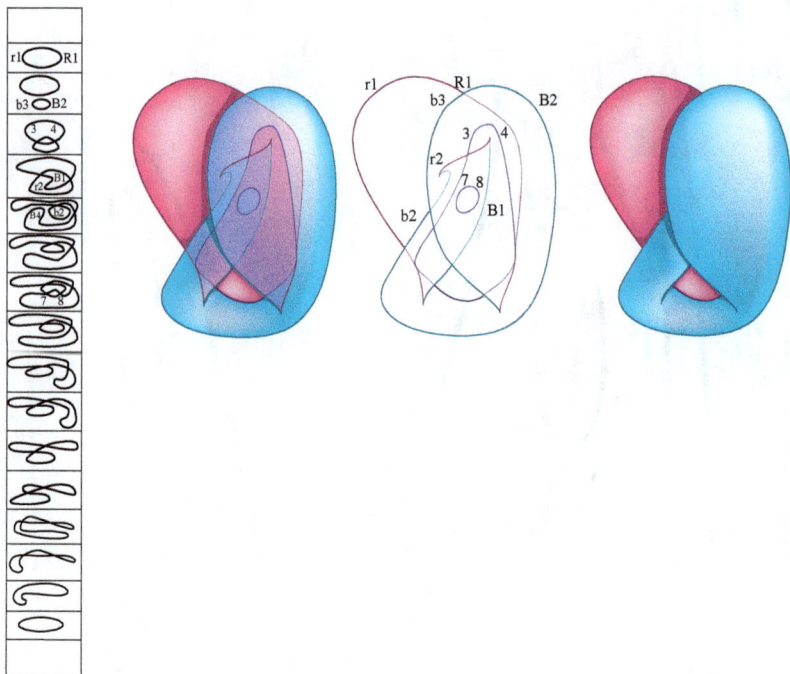

This is illustration blue # 12. In the next stage, the loop 78 of double points vanishes by a type II bubble move.

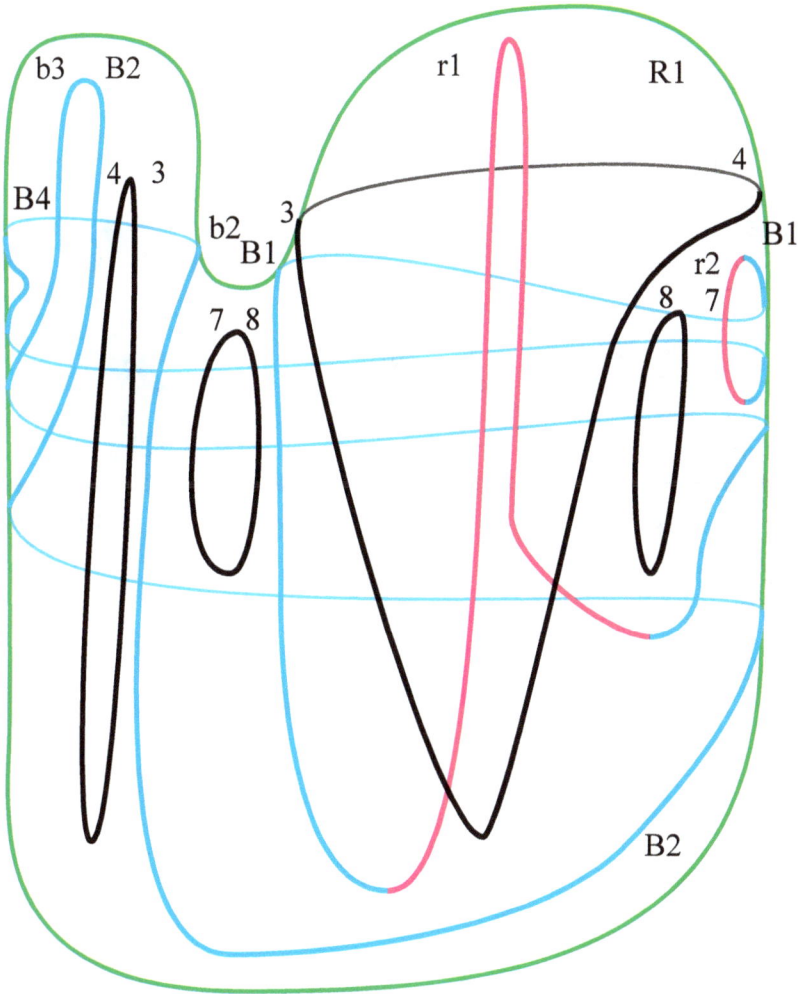

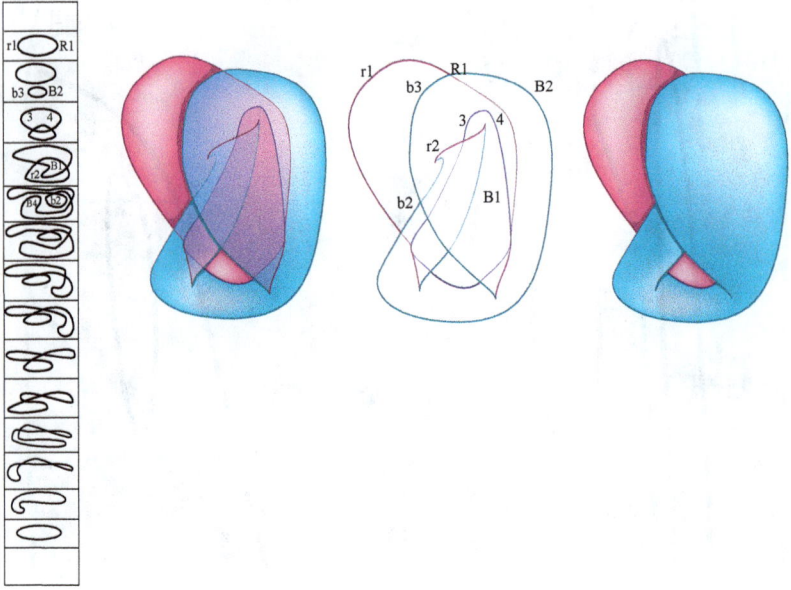

This is illustration blue # 11. In the next stage, a horizontal cusp move occurs at $r2, B4$ (which is indicated on the movie, but not the chart) thereby turning the saddle red.

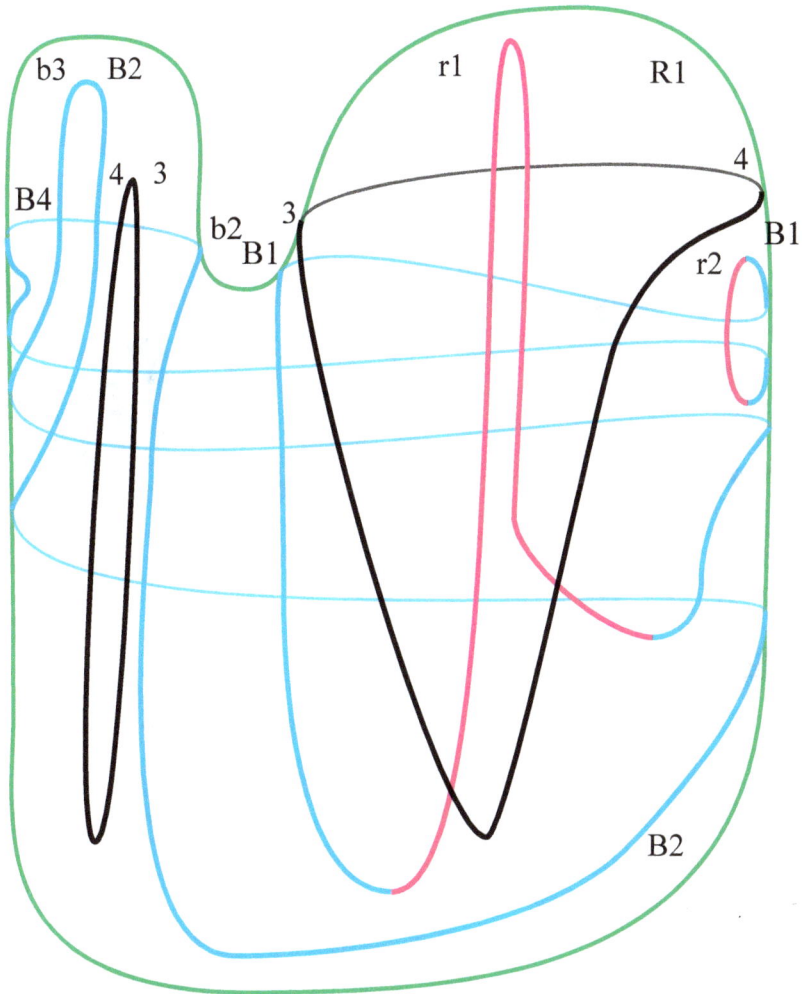

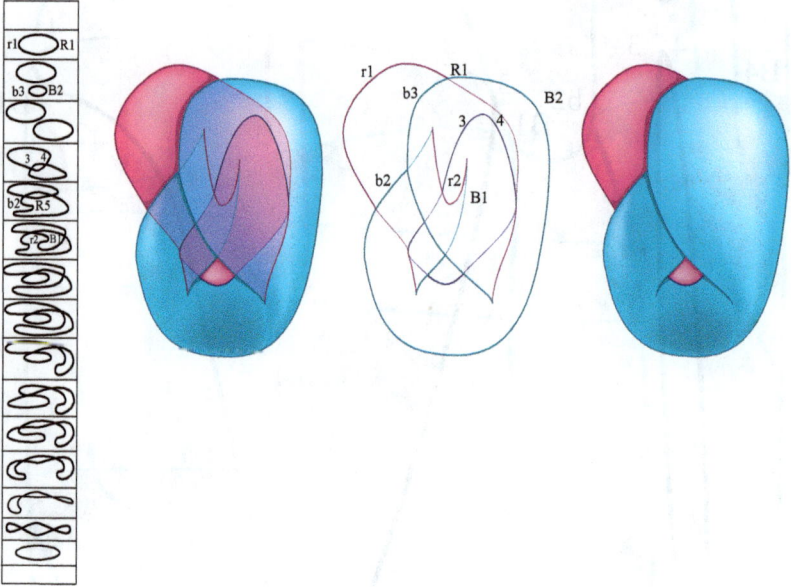

This is illustration blue # 10. In the next stage, a beak-to-beak move breaks the folds $b3$ and $R5$.

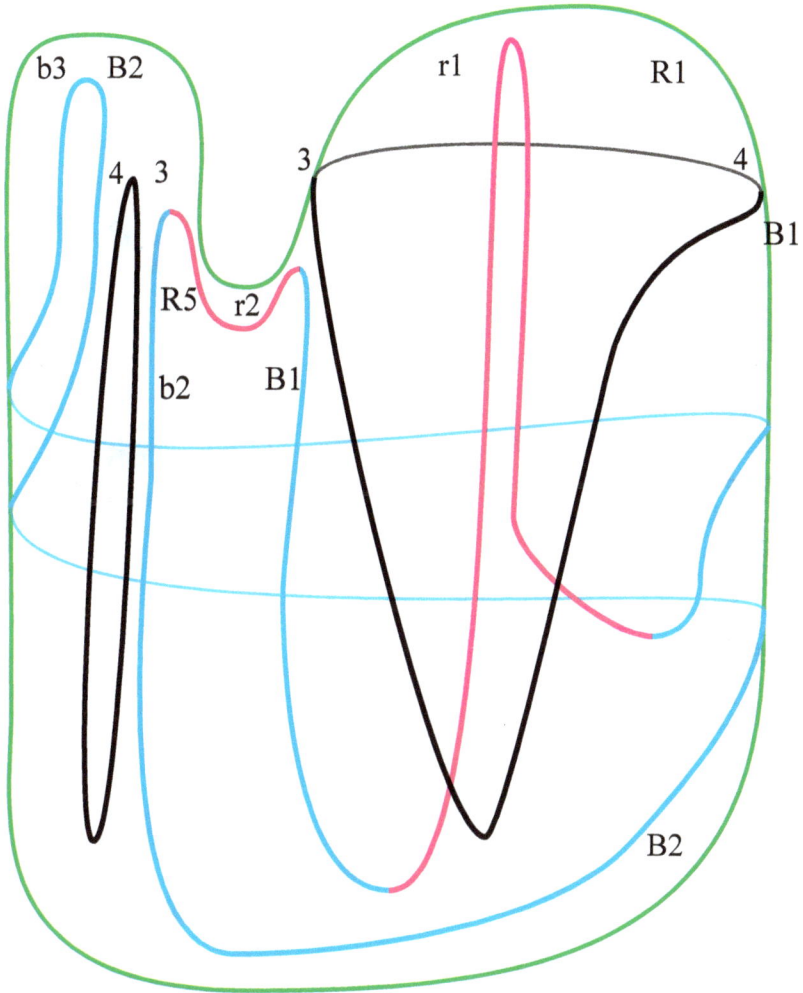

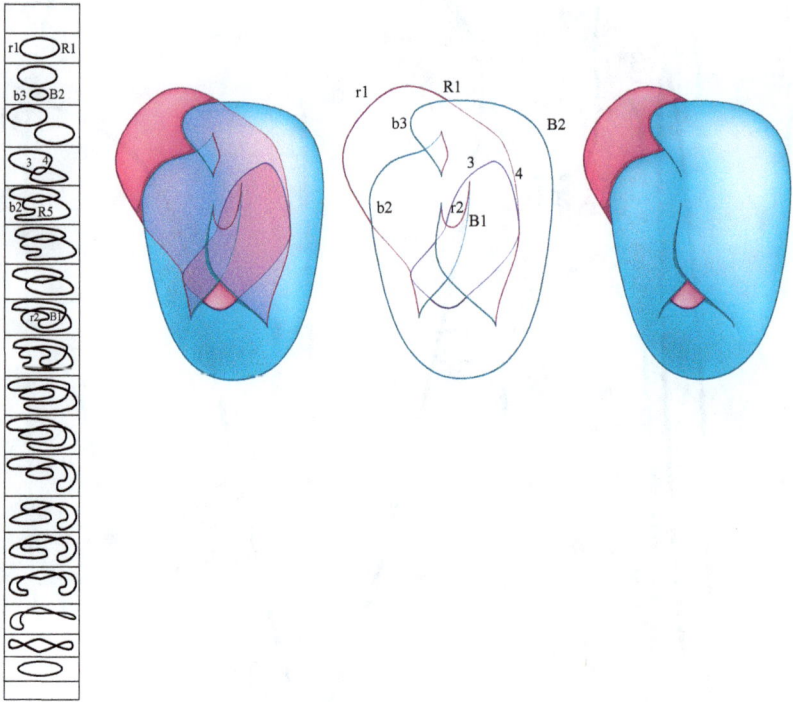

This is illustration blue # 9. In the next stage, a swallow-tail move connects the blue folds *b*2 and *b*3.

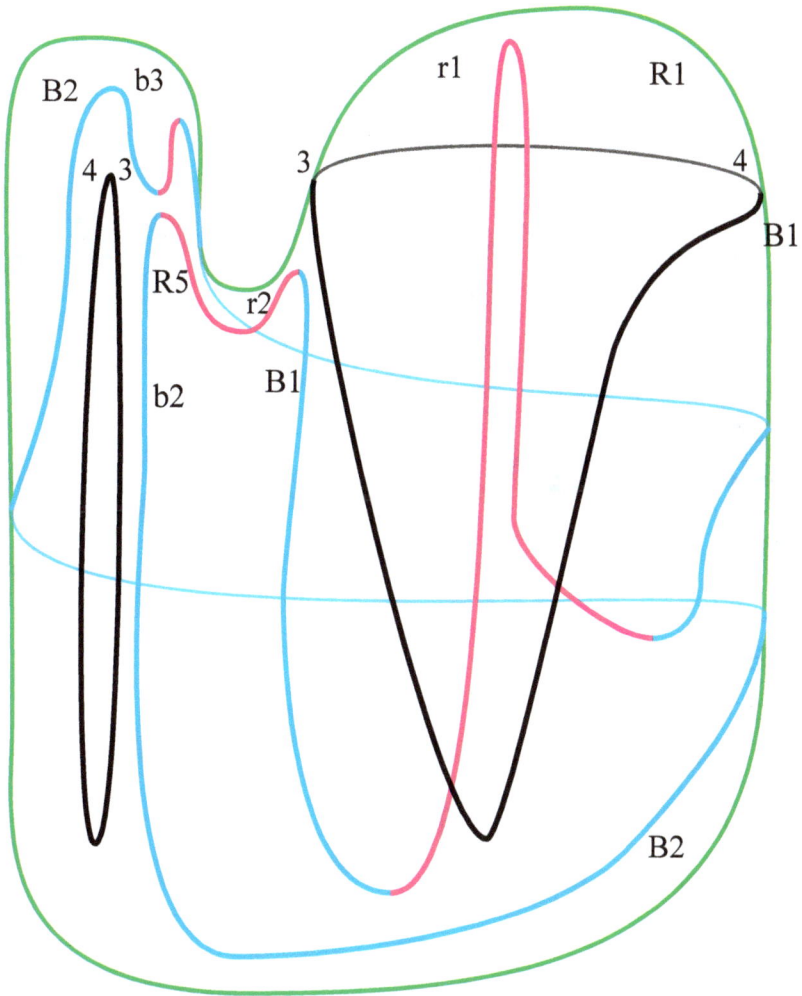

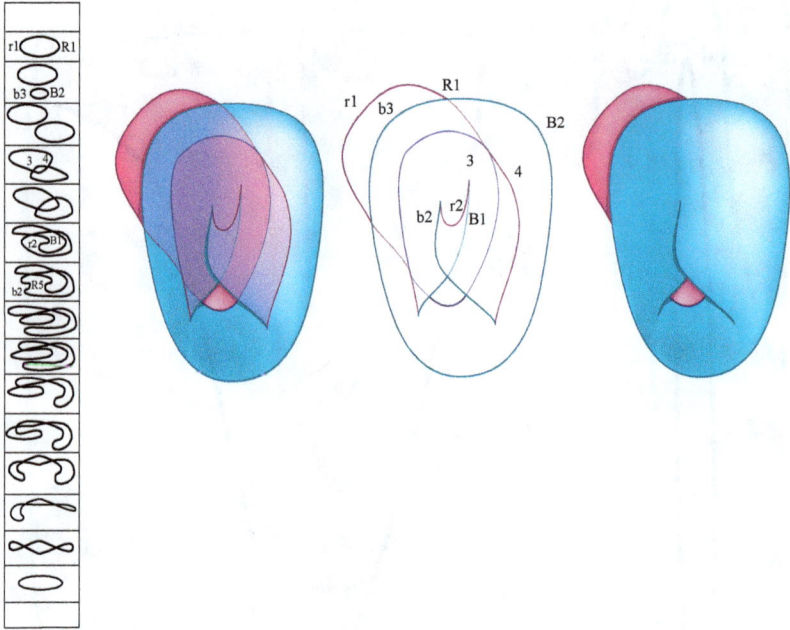

This is illustration blue # 8. In the next stage, the bounce between $R1$ and 4 moves over the death of 34 and down the left double point arc. The red death at the top moves below the blue death.

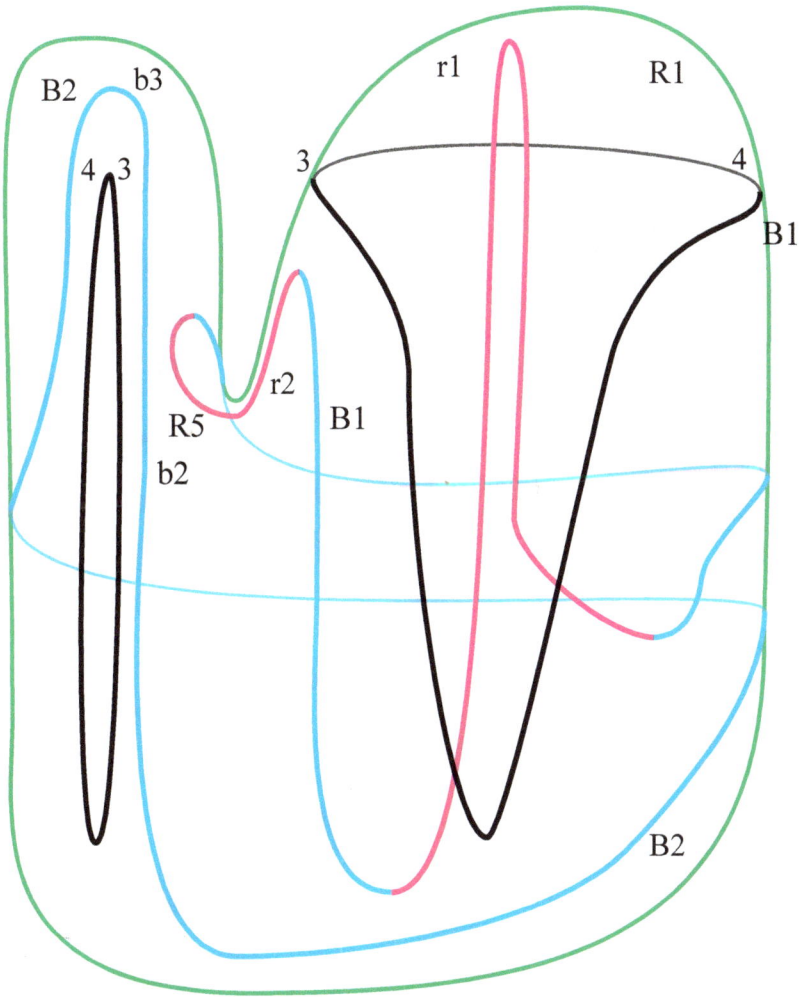

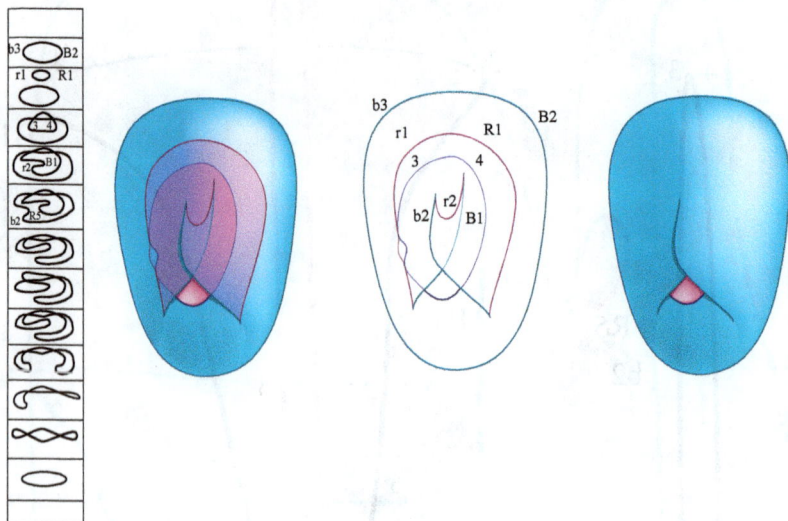

This is illustration blue # 7. In the next stage, a ψ, ψ-move removes the vestigial double bounce.

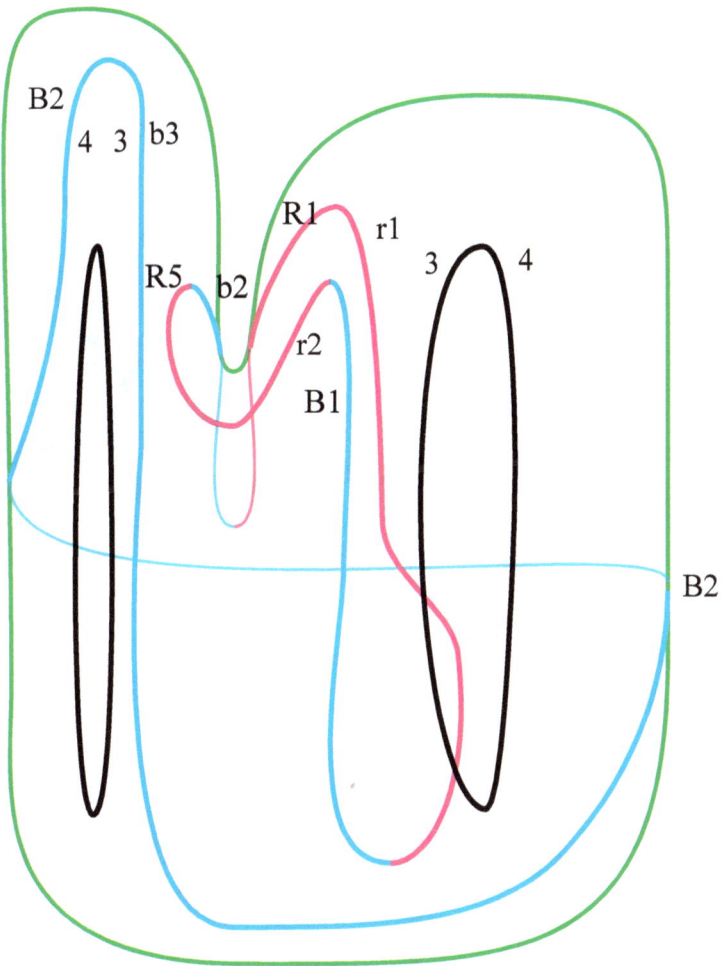

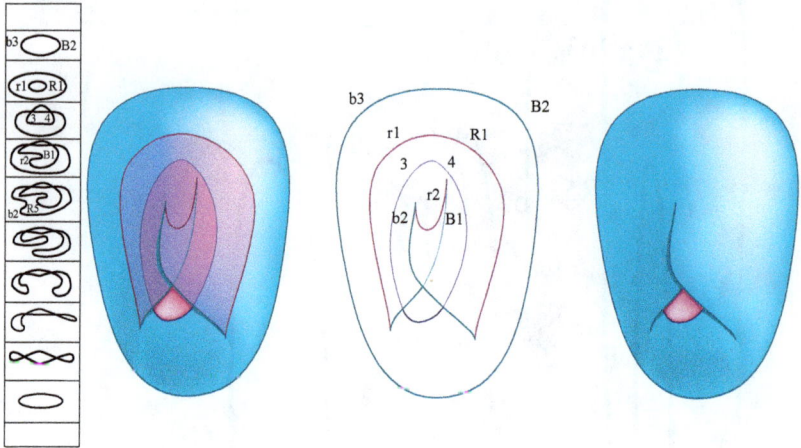

This is illustration blue # 6. In the next stage, the double curve 34 is removed by a type II bubble move.

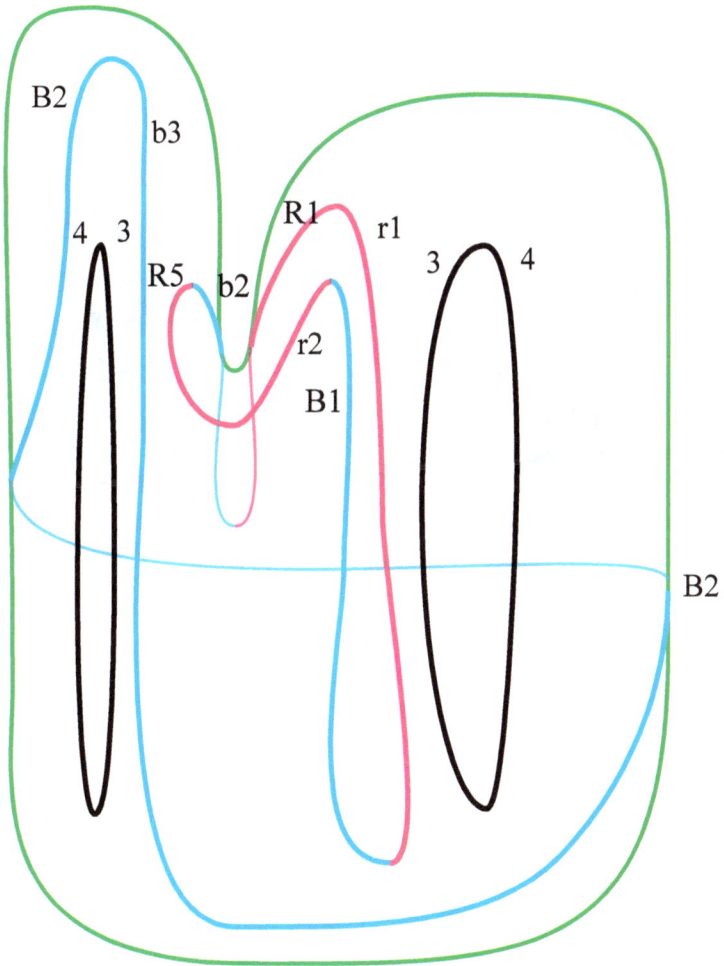

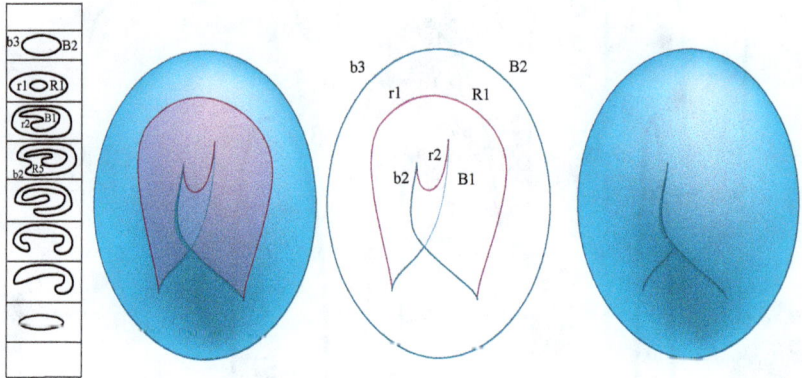

This is illustration blue # 5. In the next stage, a horizontal cusp move occurs on the left.

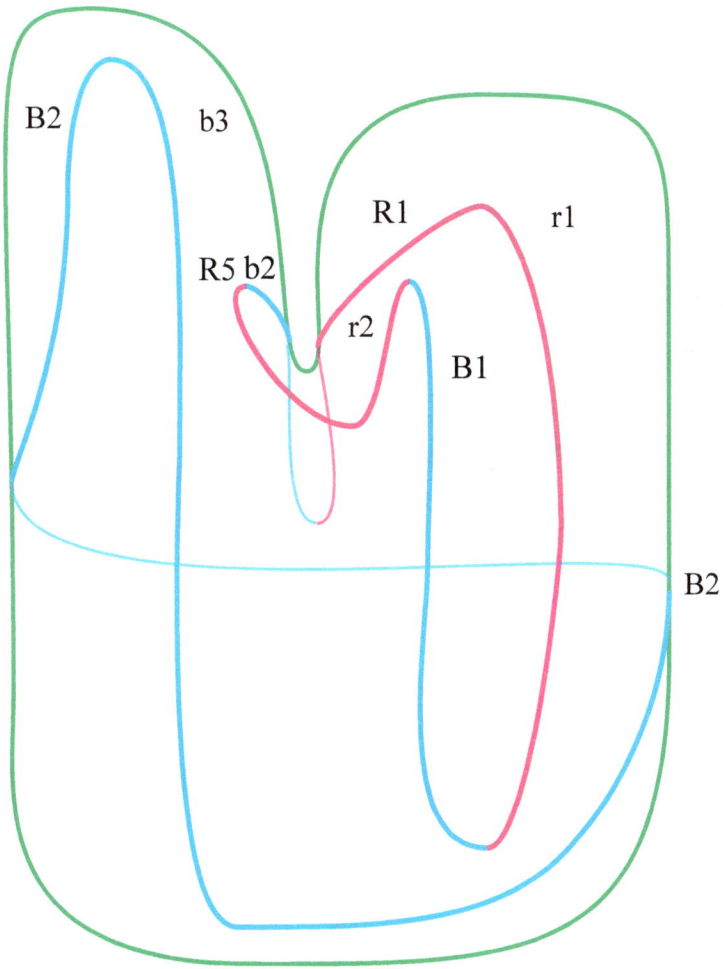

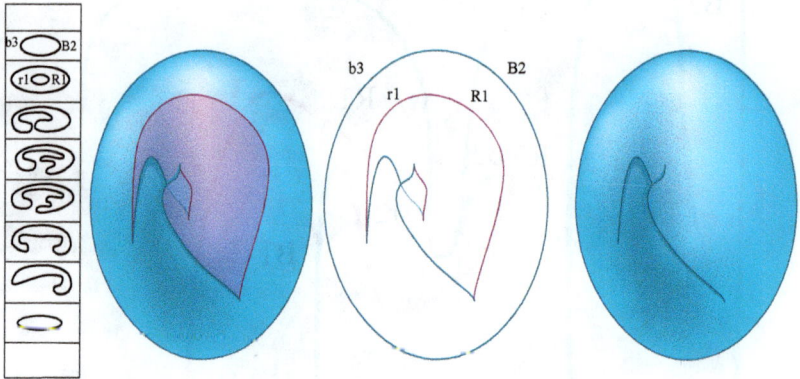

This is illustration blue # 4. In the next stage, a swallow-tail move cancels the two cusps in the middle.

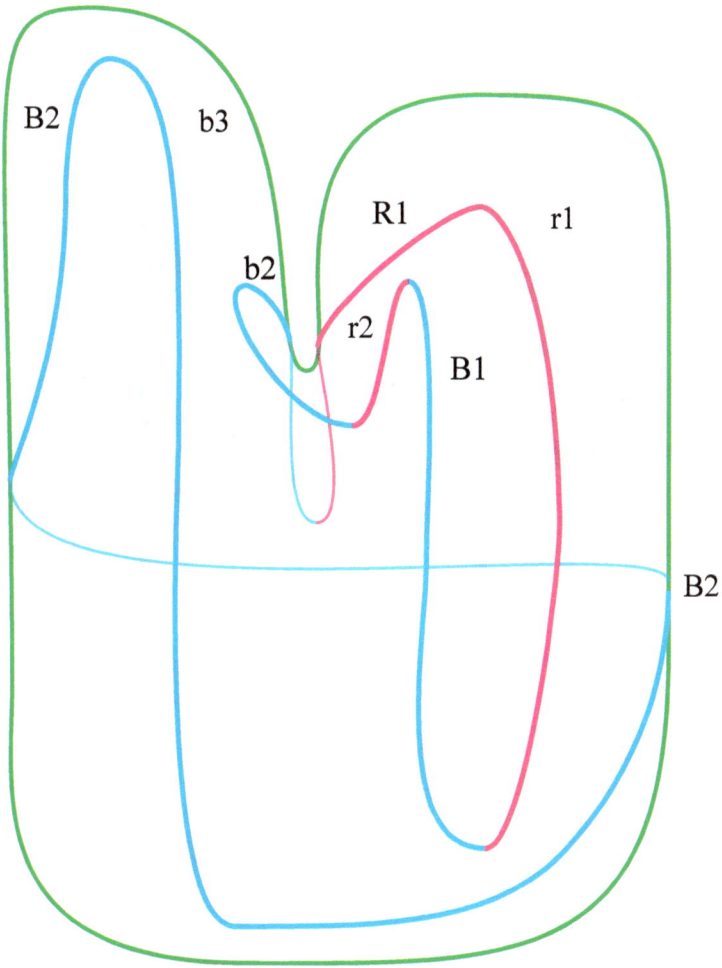

This is illustration blue # 3. In the next stage, another horizontal cusp occurs followed by a critical cancellation of a red saddle and the red maximum.

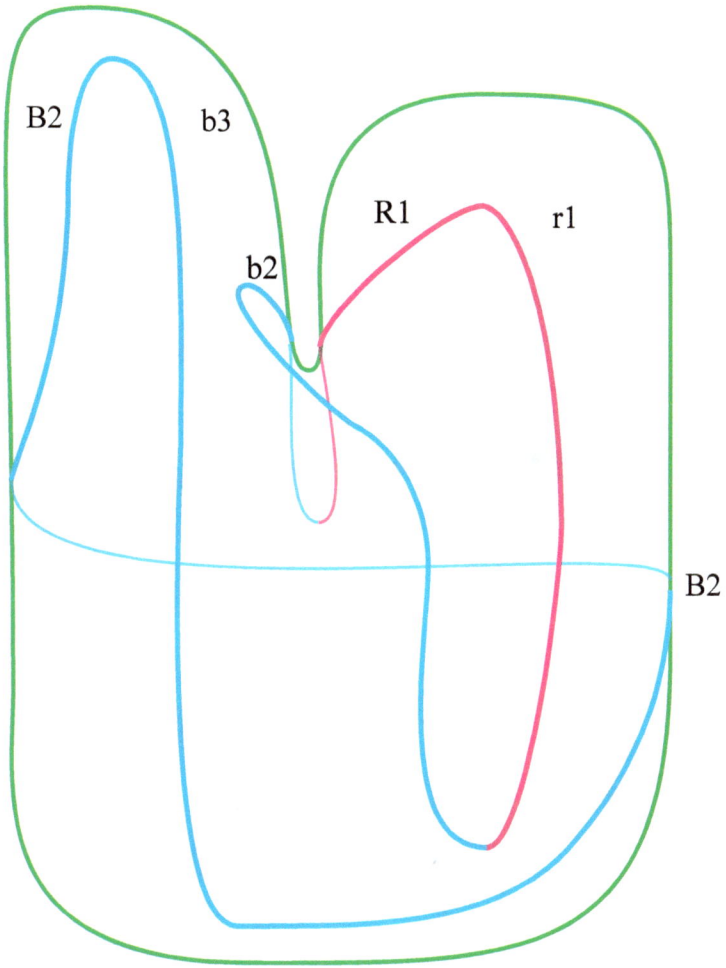

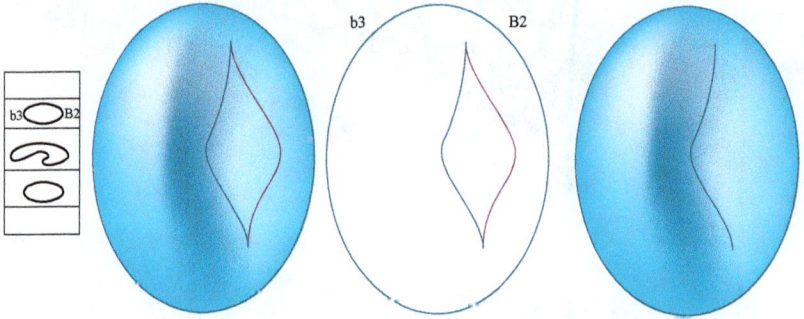

This is illustration blue # 2. In the next stage, a lips move cancels the pair of cusps in the middle.

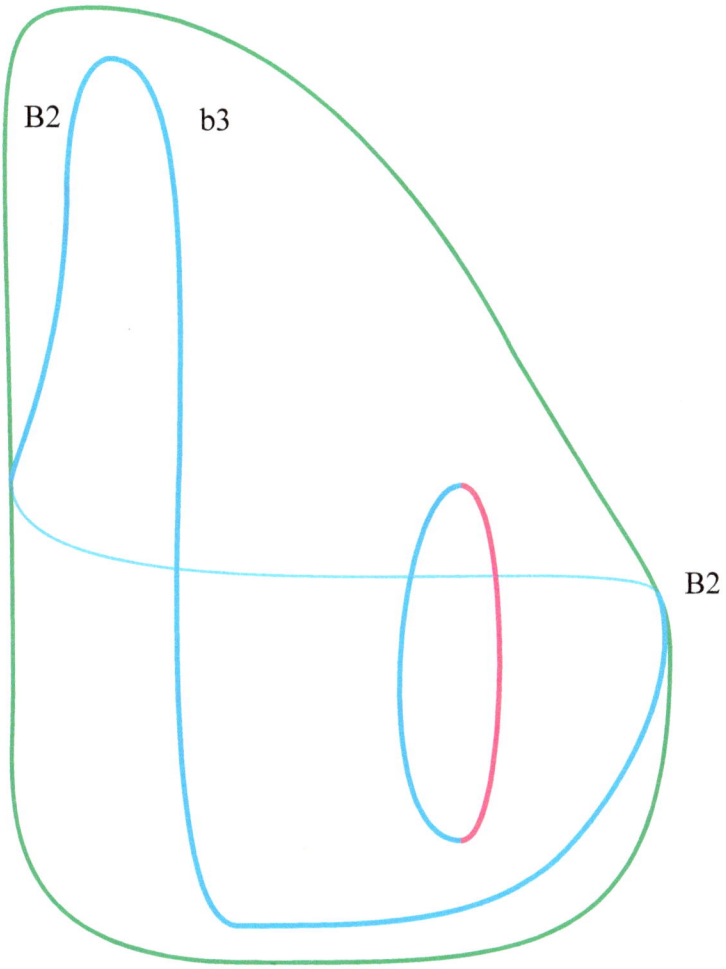

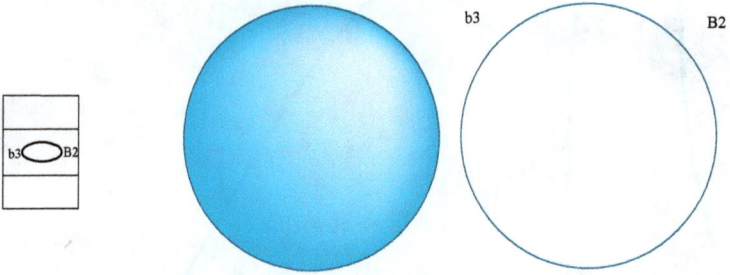

This is illustration blue # 1. The sphere is round and blue.

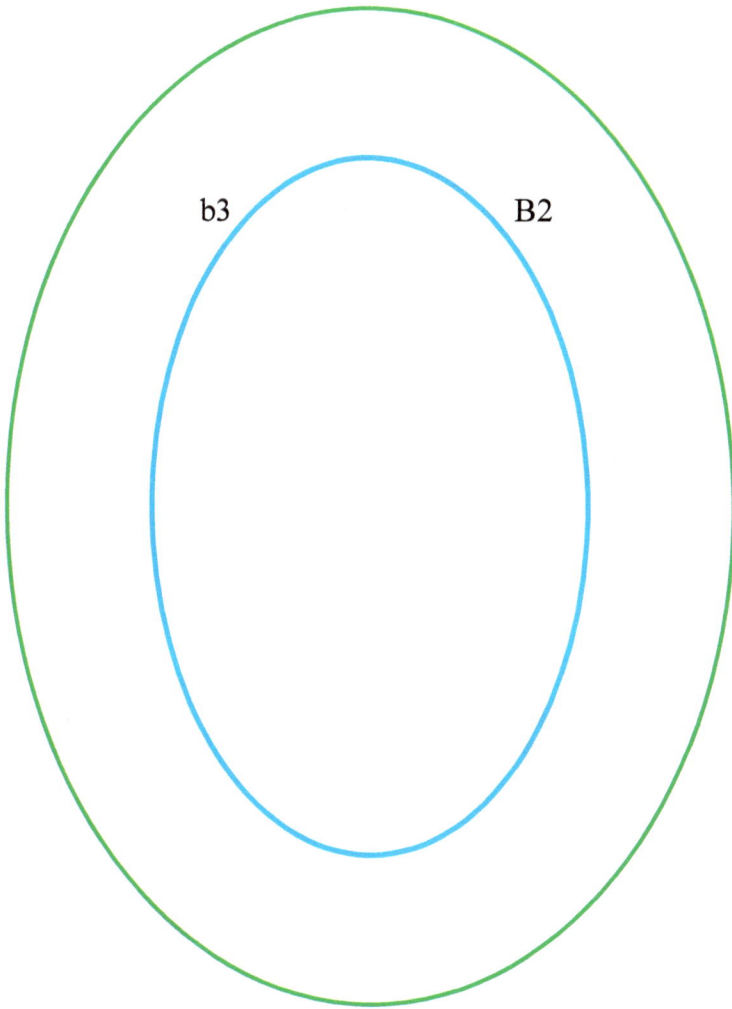

Chapter 13

The Double Point and Fold Surfaces

The double point set and the fold set of the eversion as it evolves are both surfaces immersed in space-time. Recall that a surface is *immersed* if every point has a neighborhood upon which the tangent plane provides a good approximation. So in this neighborhood the surface is embedded, but another neighborhood may overlap the surface within space-time.

The central illustrations on the even numbered pages above are conceived of as movie cross sections to these surfaces. So we can use the techniques of movies to interpolate the surfaces between successive pages. There are several ideas to discuss.

First, the double point surface is not in general position since it has triple points. In a neighborhood of the triple point arc three sheets of the double point set intersect. The situation is illustrated in Fig. 13.1. As the triple points evolve, an arc of triple points is interpolated between successive pages as the figure indicates.

Second, the critical points of the curves on the pages evolve to be critical arcs for the surfaces. Let us consider the fold set initially. The critical points are the maxima, minima, saddles, and cusps of the fold lines. A change in the fold set (such as a lips move, beak-to-beak, horizontal cusp, creation of a canceling pair of critical points, or the cancellation thereof) induces a critical point in the interpolating surface that can be seen as a critical point on the critical arc. These critical arcs envelope the time-elapsed fold surface.

An illustration of the time-elapsed fold set is indicated in Fig. 13.2 that tracks all the critical changes of the fold set from the red illustration to the blue illustration. Note that in this figure, I am breaking with top/right, bottom/left conventions. So the initial maximum of the red sphere appears on the left of the figure.

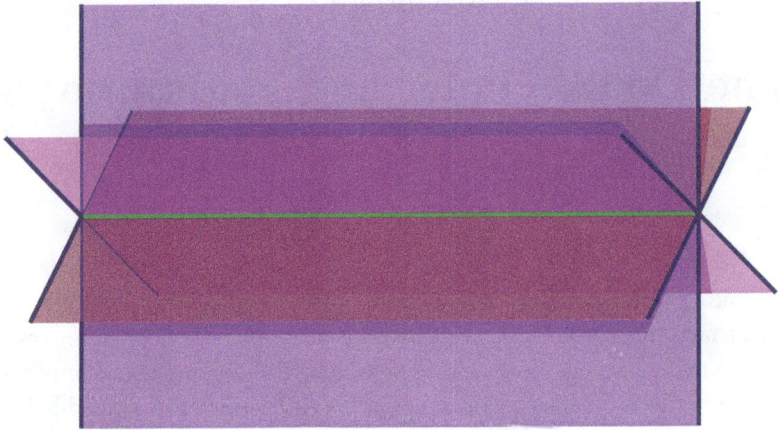

Fig. 13.1   Three sheets of double points intersect along an arc of triple points

Meanwhile, the critical points of the double point set on a single page consist of type II moves, triple points, and (should we care to keep track of them) the ψ moves upon which the double points bounce over the folds. The type II bubble, saddle, and zig-zag moves, for example, each induces a change in the critical behavior between pages.

Figure 13.4 illustrates the critical evolution of these two surfaces together. The figure represents the tool that Sarah used in rechecking the computation of her thesis. She meticulously sketched out this figure on an extended sheet of graph paper.

The color coding indicates: (1) the color of the optimum, saddle, or fold during the process, (2) the color of the cusp set during the process, (3) the triple points are colored green for no particular reason. The cusps are either red or blue, but since a cusp is a convergence of a red fold and a blue fold, the cusps are shaded different tints of purple depending on which fold is more visible.

Small words indicate the nature the critical event that occurs between pages. The relative vertical position of the critical events should roughly

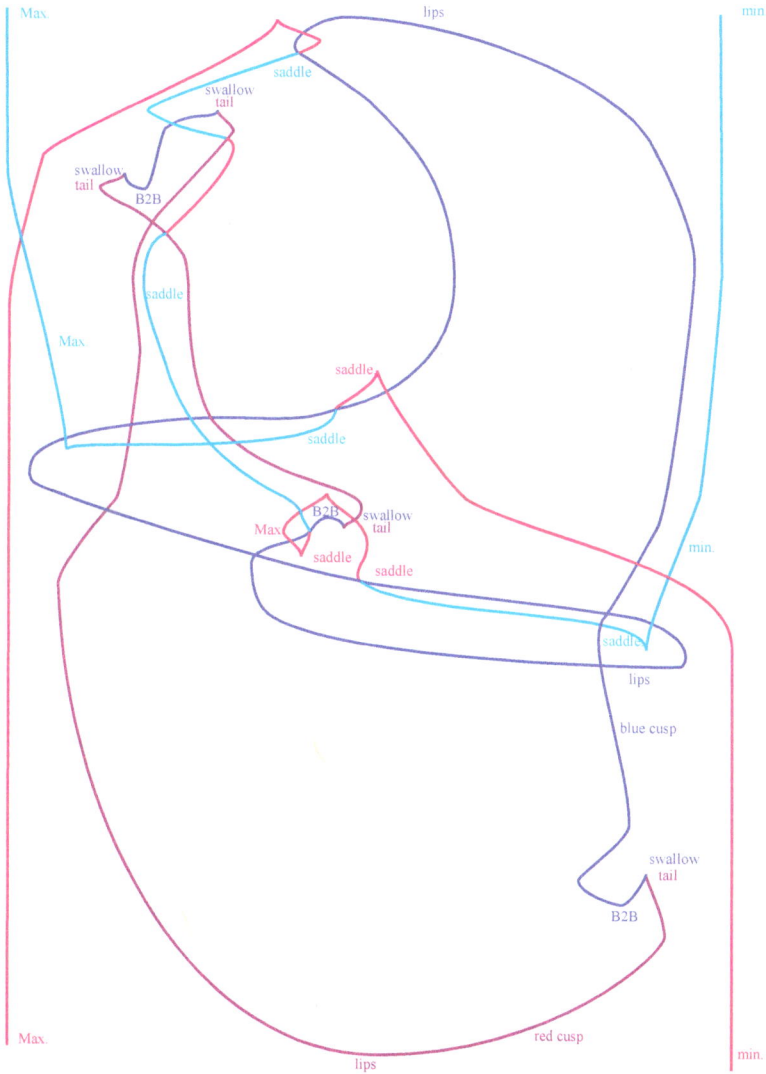

Fig. 13.2   The time-elasped view of the critical points of the fold set

correspond to the height on the page at which these incidents occur.

The next step in depicting these surfaces is to interpolate the surfaces between successive pages. For the double point surface, a rough sketch of

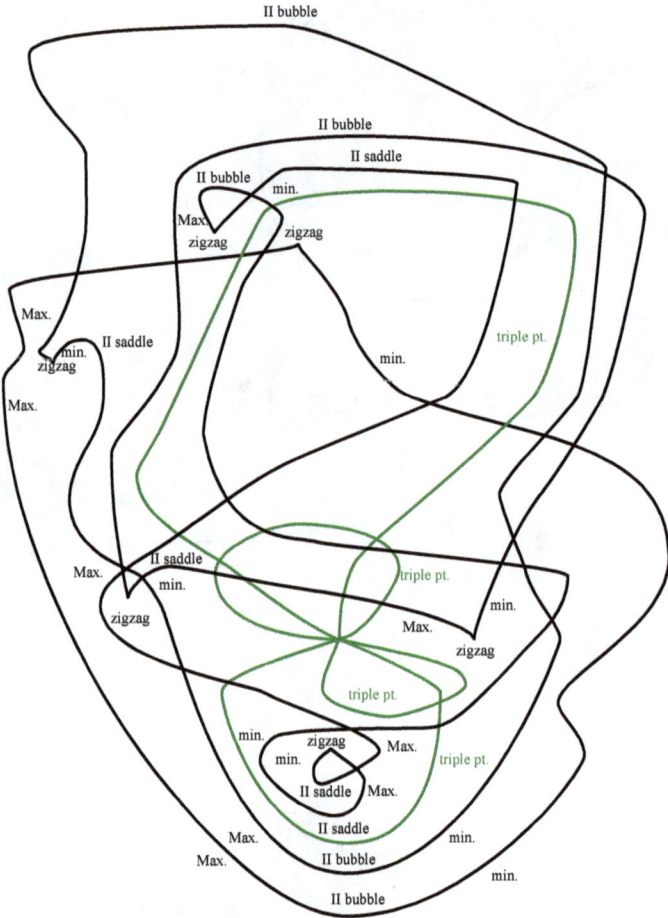

Fig. 13.3   The time-elasped view of the critical points of the double point set

this interpolation appears in Fig. 13.5 for the pages between illustration red # 28 and blue # 53. Within this illustration alone there are several sheets that overlap. Some more information could be given that indicates

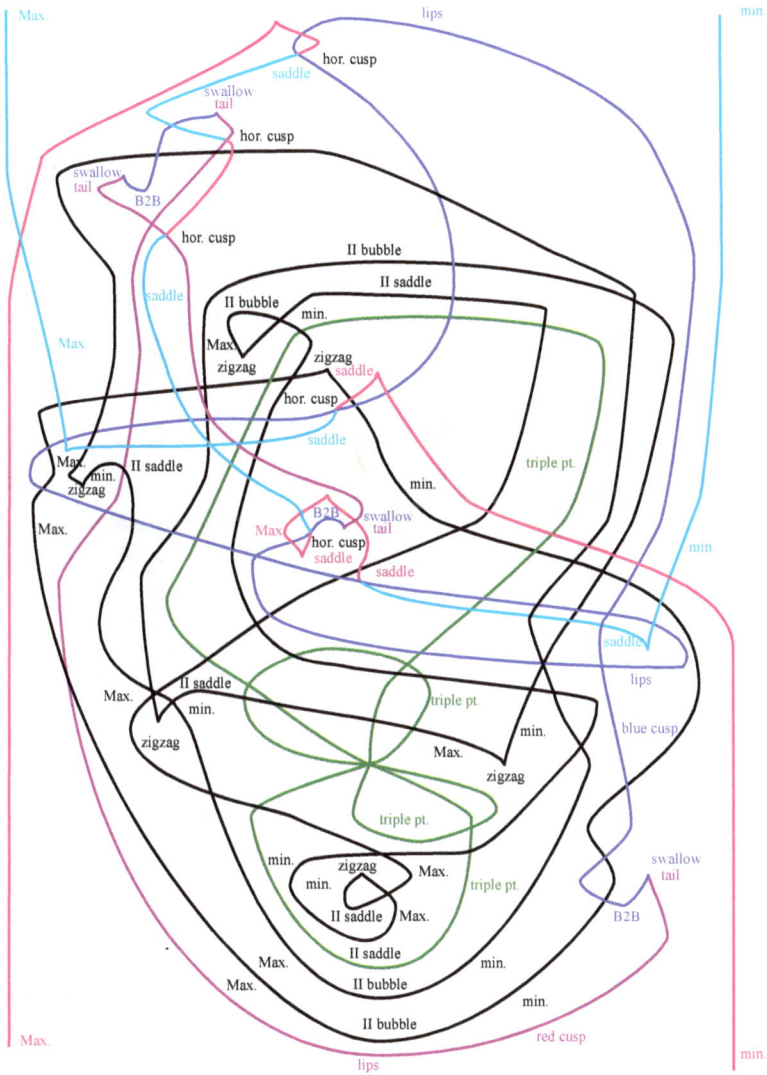

Fig. 13.4   Tracking the critical behavior of the folds, double points, and triple points

a 4th dimension which coincides with the original opacities of the double point arcs (and thus with the distance in 3-dimensional space towards the page). Figure 13.6 depicts a similar situation for the fold surfaces among the pages blue # 37 (shown on the left) through and blue # 34 (on the right).

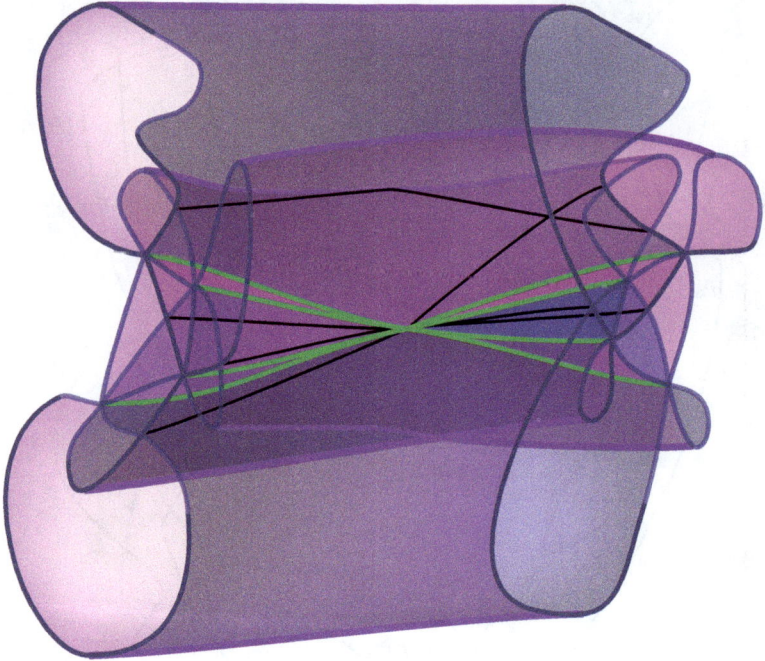

Fig. 13.5   Interpolating the double point surface on either side of the quadruple point

———

All good mathematics books contain exercises. These are items that the author can do, but might not want to. These last two illustrations are drawn at the edges of my ability. So my exercise to you, the reader, is to create an aesthetically pleasing set of illustrations that depicts the total double point surface and the fold surface for the eversion.

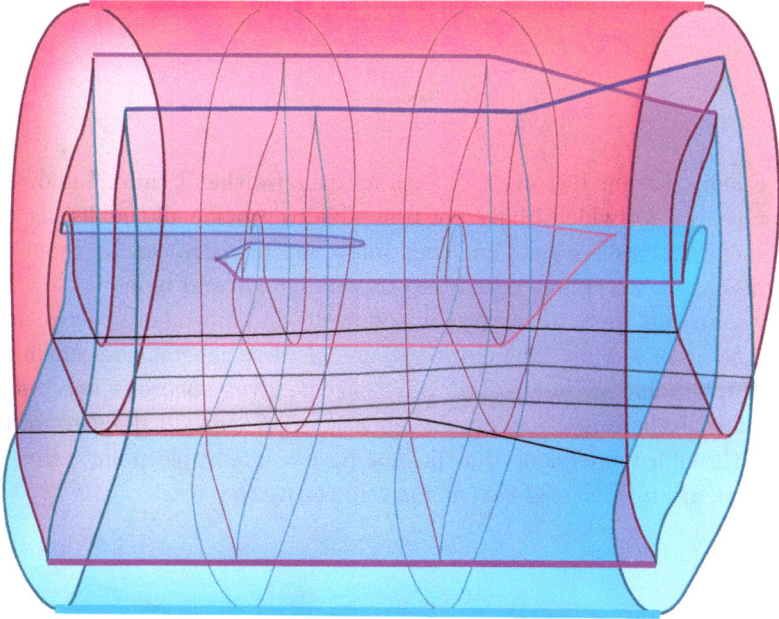

Fig. 13.6   Interpolating the fold surface between blue # 37 and blue # 34

Conclusion

Mathematicians speak of their work and the works of each other in terms of beauty. That the sphere can be turned inside-out while preserving its tangent planes is a "beautiful theorem." To witness the eversion via any of the great animations that have been produced is exhilarating. To have performed this eversion with this degree of detail has been a happy time in my life. Here each step can be examined.

However, it is not enough to reproduce another example. The example here is a very important one. For example, it describes, to the initiated, a method of mapping a very high dimensional sphere to a sphere of three smaller dimensions. The double point surface contains important information about the global eversion. The triple point set can be framed by the vectors perpendicular to the three sheets that intersect. The frame twists as it travels around the four loops of the triple point set even though the journey runs straight through the quadruple point.

———————

Finally, I want to leave you with an exercise that I think I know how to solve, but for which I have not been able to provide all the details. Any 2-dimensional sphere that has been immersed in space can be thought of as the image of a sphere that is red on the outside and blue on the inside. There are two ways to move that sphere to an embedded sphere: one is red, the other is blue. The motion to one of these spheres does not have a quadruple point in the process. Say that the given sphere is *one color or the other* is there is a way of deforming it to an embedded sphere of that color through immersions that do not have a quadruple point. *How can you look at an immersed sphere and tell which color it is?*

Bibliography

Carter, J. Scott, *How Surfaces Intersect in Space. An Introduction to Topology,* Series on Knots and Everything, 2. World Scientific Publishing Co., Inc., River Edge, NJ, (2nd Edition 1995).

Carter, J. Scott; Saito, Masahico, "Reidemeister Moves for Surface Isotopies and Their Interpretation as Moves to Movies," J. Knot Theory Ramifications 2 (1993), no. 3, 251–284.

Carter, J. Scott; Rieger, Joachim H.; Saito, Masahico, "A Combinatorial Description of Knotted Surfaces and Their Isotopies," Adv. Math. 127 (1997), no. 1, 1–51.

Francis, George; Sullivan, John M.; Kusner, Rob B.; Brakke, Ken A.; Hartman, Chris; Chappell, Glenn, "The Minimax Sphere Eversion. Visualization and Mathematics" (Berlin-Dahlem, 1995), 3–20, Springer, Berlin, 1997.

Max, Nelson; Banchoff, Thomas, "Every Sphere Eversion Has a Quadruple Point," Contributions to analysis and geometry (Baltimore, Md., 1980), pp. 191–209, Johns Hopkins Univ. Press, Baltimore, Md., 1981.

Morin, Bernard; Petit, Jean-Pierre, "Le Retournement de la Sphère. In Les Progrès des Mathématiques, pp 32-45. Pour la Science/Belin, Paris, 1980.

Phillips, Anthony, "Turning a Sphere Inside Out," Sci. Amer. 214 (1966), 112–120.

Smale, Steven, "A classification of immersions of the two-sphere," Trans. Amer. Math. Soc. 90 (1959), 281–290.

Sullivan, John M.; Francis, George; Levy, Stuart, "The Optiverse," In H.-C. Hege and K. Polthier, eds, *VideoMath Festival at ICM'98*, Springer, 1998. Video (7 min). See also `http://video.google.com/videoplay?docid=-761214833095493063`
or `http://www.youtube.com/watch?v=cdMLLmlS4Dc`

Thurston, Nathaniel, *et al.* `http://video.google.com/videoplay?docid=-6626464599825291409`

Index

SERIES ON KNOTS AND EVERYTHING

Editor-in-charge: Louis H. Kauffman *(Univ. of Illinois, Chicago)*

The Series on Knots and Everything: is a book series polarized around the theory of knots. Volume 1 in the series is Louis H Kauffman's Knots and Physics.

One purpose of this series is to continue the exploration of many of the themes indicated in Volume 1. These themes reach out beyond knot theory into physics, mathematics, logic, linguistics, philosophy, biology and practical experience. All of these outreaches have relations with knot theory when knot theory is regarded as a pivot or meeting place for apparently separate ideas. Knots act as such a pivotal place. We do not fully understand why this is so. The series represents stages in the exploration of this nexus.

Details of the titles in this series to date give a picture of the enterprise.

*The complete list of the published volumes in the series, can also be found at
http://www.worldscibooks.com/series/skae_series.shtml

www.ingramcontent.com/pod-product-compliance
Lightning Source LLC
Chambersburg PA
CBHW050545190326
41458CB00007B/1924